# Violent Universe

Nigel Calder

# Violent Universe

An Eyewitness Account
of the New Astronomy

New York / The Viking Press

# Contents

*Colour illustrations are between pp. 40–1 and pp. 136–7.*

*Author's note*
The author acknowledges the opportunities to visit observatories abroad afforded him by the BBC and Public Broadcast Laboratory and by the Commonwealth Scientific and Industrial Research Organization of Australia. Most of the information and nearly all the perspectives in this book were obtained in conversations with about 100 astronomers. Details have been checked in papers appearing in learned journals (especially in *Nature, Science, Astrophysical Journal* and *Physical Review Letters*) and in more general accounts (especially in *Scientific American, New Scientist, Science Journal* and *Science Year*). Much of the subject matter is too fresh to have appeared in up-to-date form in other books for the general reader, but anyone wishing to read further may consult the following, which have also been of assistance to the author:

D. W. Sciama, *The Unity of the Universe* (Faber & Faber, 1954).
George Gamow, *A Star Called the Sun* (Macmillan, 1964).
F. Graham Smith, *Radio Astronomy* (Penguin Books, 1966).
*Larousse Encyclopaedia of Astronomy* (Paul Hamlyn, 1968).
*The New Universe* by *Science Journal* (Iliffe, 1968).

*Note on units*
*Billion* is used in the American sense of 1,000 million.
Temperatures are given in centigrade degrees absolute (K).

*Illustrations*
Sky photos are acknowledged in their own captions. The other photos were supplied as follows: page 13 (bottom), National Radio Astronomy Observatory; 14, Arecibo Ionosphere Observatory, Cornell University; 15 (left), Homestake Mining Co.; 16, USIS; 23, NASA; 31, Bristol University; 38, Chatterton Astrophysics Dept; 40, California Institute of Technology; 44, National Gallery; 51, A. C. K. Ware Ltd; 53, NASA; 59 (right), COI (Crown copyright); 65, Radio Times Hulton Picture Library; 66 (bottom), Brookhaven National Laboratory; 72, Kitt Peak National Observatory; 87, Lick Observatory; 103 (right), NRC, Canada; 111, Fox Photos; 122, SRC; 124, COI (Crown copyright); 127 (top), SRC; 127 (bottom), COI (Crown copyright); 140, 147, 148, MIT Lincoln Laboratory; 146, Radio Times Hulton Picture Library; 153 (top & right), NASA; 153 (bottom), American Science and Engineering Inc.; 154, 155, Paris Observatory; colour: facing p. 41 (top), Bristol University; facing p. 137, Bell Telephone Labs. All the remaining photos are by Philip Daly.

The diagrams were drawn by Jill McDonald

*Golden age*

The big discoveries raise questions that make astronomers work feverishly and argue with an agitation that verges on rudeness. One radio astronomer reports that he has lost fourteen pounds avoirdupois in following up disclosures of the pulsars. An optical astronomer confesses that he most wants to find a quasar travelling at enormous speed, to settle the hash of those who quarrel with his interpretation of these strange objects. A distinguished theorist presents his views on this subject and a learned audience giggles. Each new result is greeted with glee by one school of theorists or another, while their rivals rush to shore up their shaken structure of ideas.

This is astronomy today: a giddy intellectual game for great telescopes and great minds. What is at stake is nothing less than the imminent portrayal of a new and more vivid picture of the universe we inhabit, as astronomers routinely investigate cataclysms in the sky. It turns out that we live in a relatively peaceful suburb of a quiet galaxy of stars, while all around us, far away in space, events of unimaginable violence occur. There are objects so disorderly that they seem to violate even the laws of physics, as patiently elucidated and verified by ten generations of scientists on Earth.

Pulsars? Quasars? Galaxies? Do I presume that the reader already knows what these objects are? Certainly not, because the astronomers don't know! They are names given to phenomena observed in telescopes. They can be described but not yet explained. Pulsars, the discovery of which was announced by Cambridge radio astronomers in 1968, are 'pulsating radio sources', stars that broadcast great bursts of radio energy with the regularity of a ticking clock. They may be rapidly spinning stars, so crushed by their own gravity that a million tons of matter in them would fit in the volume of a pea. Quasars are 'quasi-stellar radio sources' or 'quasi-stellar objects', discovered in the early 1960s. As their name suggests, they look like nearby stars, but they are far away and very bright – fantastic concentrations of energy that defy explanation.

The reader who recognises 'galaxies' as a familiar term – for those vast collections of stars scattered like ships in the ocean of space – may be surprised to see them bracketed with the pulsars and the quasars, as unknown quantities. The reason is that ideas are being drastically revised, since an uproar of radio noise from the sky makes it plain that galaxies are something more than the sum of their component stars. Evidence now piles up, showing that galaxies can undergo violent changes, amounting in some cases to vast central explosions. The energy then radiated, and measured in the telescopes, is far greater than the normal stars themselves could produce. An obvious question is whether quasars are exploding galaxies.

Current astronomy concerns itself with birthdays and doomsdays – for the Earth, for the Sun and other stars, for the galaxies and for the mighty universe itself. Although we are seeing a story unfold, with dénouements still to come, the astronomers already have fascinating narratives of nature. Aided by a world-wide watch on the Sun, and by rockets and satellites, they have shown that our private star, an unchanging ball of fire to the casual eye, is a scene of storms and explosions. It is also the source of a solar wind of atomic particles that blows continuously outwards through the system of planets. The astronomers can tell how the Earth's cargo of atoms, from which we ourselves are made, came into existence in a process of nuclear assembly in the interior of stars long since exploded. They can explain why iron is common and gold is rare. And they predict a day when the ageing Sun will swell and swallow up the earth. Some of them forecast that, much later, the whole universe will annihilate itself.

For a long time the Earth was the centre of the universe; after Copernicus the Sun was; fifty years ago the universe was just *the* Galaxy, the Milky Way, in which the Sun was a modest member among 100,000 million stars. If that seemed world enough for anyone, it has since turned out to be a great understatement, because we now know that there are thousands of millions of galaxies. And radio telescopes pick up a strange whisper from empty sky. Many astronomers believe it to be the echo of a Big Bang about 10,000 million years ago. when all the matter of the universe was gathered in one place and was then blasted into an expanding cloud, which now can be seen in the form of galaxies moving away from us, in whatever direction we care to look. Work in progress will show us whether the Big Bang really happened; it may give us, at long last, a reliable impression of our own place in space and time.

Each science has its heyday. The 1930s and 1940s were dramatic and fateful years for the atomic physicists. The 1950s brought a revolution in biology, by way of the investigation of the large molecules of life. The 1960s, and especially these closing years of the decade, are a golden age of astronomy, likely to rank in future histories with the period in the early 17th century when Galileo first turned a telescope towards the sky, and Kepler elucidated the motions of the planets. Will the 1970s bring a grand synthesis of new laws of nature, such as Newton gave? It looks as likely as not.

Looking around, astronomy is a world-wide activity, but mainly of the rich, science-based countries. Its citadels are at present in England, New England, California and New South Wales. Contributions smaller in scale but high in quality come from several other centres, most notably the Netherlands, France, Canada and Japan.

The British Broadcasting Corporation and the Public Broadcast Laboratory of New York considered a joint enterprise to make an unusually ambitious programme on a scientific subject. With a generous budget, the viewer could be taken anywhere in the world, to any laboratories which would help tell the story. As an evening's intellectual entertainment, it was to be outside the common run of science programmes for television. Astronomy, in its pregnant state, emerged as the best possible subject for such a programme. Philip Daly of the BBC was named producer and he and I met at London Airport in September 1968 – he fresh from a programme on immunology, I from a book on science and politics. To gather information and film for the programme we travelled, between us, distances equivalent to several times round the globe. This book, like the programme, is based on these and earlier travels, and on conversations with a hundred astronomers and physicists of twelve countries.

*Tsutomu Seki of Kochi, Japan. With his small rooftop observatory, this young guitar instructor keeps alive a venerable branch of astronomy – the search for new comets. He explores with the binoculars (right) and uses the telescope (left) to determine the trajectory of a new-found comet.*

*The Ikeya-Seki comet, which Seki spotted simultaneously with another Japanese amateur in 1965, was his most celebrated discovery. (Lick Observatory)*

*Opposite:*
*Parkes, New South Wales.*
*The 210-foot radio telescope of the Commonwealth Scientific and Industrial Research Organization has been responsible for several important contributions to astronomy, including the pinpointing of the quasar 3C273 in 1962. The 60-foot telescope in the foreground runs on rails and can be operated jointly with the big dish at a chosen spacing.*

Daly and I had to visit the great American observatories. We also checked what was going on at the principal British centres, Cambridge and Manchester Universities, and the Royal Observatories at Herstmonceux and Edinburgh. We kept in touch with developments at the main locations in Australia: Parkes, Culgoora, Molonglo and Mount Stromlo. We were to go among the bears and beavers, to the lakeside forest of Algonquin Park in Canada, where the radio astronomers have pioneered the use of tape recorders to link telescopes thousands of miles apart. In Tokyo, we wanted to talk with professional astronomers and physicists, and then move on to the Japanese island of Shikoku, to meet the young guitar instructor, Tsutomu Seki, who has discovered five new comets with a rooftop telescope. We also cast widely for theoretical ideas: in Cambridge, England, and Cambridge, Massachusetts; London and Sussex; Cornell, Princeton and Caltech – all these were on our itinerary. I had attended a conference in Trieste, where leading astrophysicists from several countries rehearsed the state of their knowledge for their fellow physicists.

What about the USSR? When we were planning the BBC-PBL programme, a check-list of Soviet observatories and astronomers and lines of research did not add up to an obligation to film there. The

*Joseph Shklovsky. A Soviet
theoretical astronomer who has
made important contributions to
the explanation of radio sources
in the sky.*

Georgian, Victor Ambartsumian, working at Yerevan in Armenia, is an astrophysicist of world rank. He made several key discoveries about the formation of stars; even more significantly, it was Ambartsumian who first pointed to the bright nucleus of a galaxy, as a region where upheavals could happen. But from the Caucasian spring, great streams of research now flow in California. Since the time of Peter the Great, mathematics and theoretical physics have been a Russian speciality and the tradition is continued by men such as Joseph Shklovsky, with his theories of radio sources in the sky. Yet recent observational results in the USSR are not impressive compared with, say, those in Australia, although there are important observatories near Leningrad and Moscow and in the Crimea. The world's biggest optical telescope – with a mirror 236 inches in diameter – is now under construction; anyone making a comparable programme about astronomy in 1979 will surely be obliged to visit it, on its mountaintop in the Caucasus.

We headed West. Nature in a big, structurally simple continent conspires with the works of ambitious men to lend a special grandeur to the scenes of American astronomy. One sunny day, flying from San Fransisco to Tucson *via* Los Angeles over the coastal range of California and into the desert, I picked out the great optical observatories that make the south-western States the eye of the Earth. The white, sunproof domes lie like the eggs of some giant species of bird, on Mount Hamilton and Mount Wilson, Palomar Mountain and Kitt Peak. On Haystack Hill, north of Boston, nestles another big dome, but this time housing a big radio dish, borrowing the anti-weather technology of the early-warning radars. And in the quiet wooded countryside of West Virginia, where the fall runs riot with red and yellow, the US National Radio Astronomy Observatory has taken refuge from the radio noise of civilisation, and uses a variety of telescopes to tune into the radio noise of the sky.

Papago Indians, Puerto Ricans and Dakotan goldminers play host to American astronomers. Kitt Peak in Arizona stands in the Papago reservation, and months of diplomacy went into negotiating the lease of a sacred mountain to 'the men with long eyes', as the Indians call their tenants. Not far away is Baboquivari, a crag pointing like an arrowhead at the sky; according to the chanted legends of the Papagos it is around this mountain that the Sun and stars revolve. Today, among its range of optical telescopes, the Kitt Peak National Observatory boasts the world's biggest instrument dedicated to looking at the Sun. Above ground a huge leaning tube, propped on a tower, dominates one skyline. The tube is but the entrance of a 500-foot tunnel for sunbeams that runs far into the mountainside: at the end, a 60-inch mirror casts back the rays to an observing room where a 33-inch image of the Sun falls on a round white table. The table looks decep-

tively ordinary; in fact, beneath it, a huge vacuum tube, diving 65 feet into the rocks, carries the spectroscope that breaks up the sunlight into component colours and lines.

Most spectacular of all the world's observatories is the radio telescope at Arecibo, near the northern coast of Puerto Rico. One of the natural craters in the *karst*, the curious dimpled landscape formed by the surface of the island caving in, has been rounded off and converted into a 1000-foot radio dish. A metal net strung across the valley floor collects radio waves from the sky overhead and focuses them to a

*The Lick Observatory, on Mount Hamilton in northern California.*

*Green Bank, West Virginia. The US National Radio Astronomy Observatory possesses a variety of telescopes including a partially steerable 300-foot dish, a fully steerable 140-foot dish, and smaller instruments.*

*Arecibo, Puerto Rico. The huge telescope of the Arecibo Ionospheric Observatory, built and operated by American scientists, consists of a 1000-foot reflecting mesh hung in a valley floor, and aerial feeds carried on a platform suspended on cables from three towers. (See also colour illustrations facing p. 136.)*

point 435 feet in the air. There, a variety of aerials wait for them, hanging from a platform which is itself suspended on cables from three tall towers around the crater. You need a good head for heights to venture out to the platform along the catwalk with nothing but a light aluminium mesh and 400 feet of air, between yourself and the ground. But you are really quite safe, and the view is worth the vertigo, to see what should figure on any short-list of man-made wonders of the modern world.

Most paradoxical is the observatory deep underground, in a cavern in the Homestake gold mine, in South Dakota. In the 1870s the gold rush hit the Black Hills. In Deadwood Gulch, Bobtail Creek and Gold Run Creek, feverish men searched and fought for the precious metal. Today, Wild Bill Hickok and Calamity Jane lie buried at Deadwood and all the surface gold is long since gone. But at Lead the Homestake gold mine still follows the ore deep into the ground, and the successors of the original campers conduct a large-scale operation as shift-work miners. Tourists come in search of reminders of the old frontier days in the land of the Sioux. It is an improbable place for astronomy.

But there Raymond Davis, a chemist if you please, uses a tank of cleaning fluid to observe what goes on inside the Sun. His bizarre astronomical instrument, without lenses or mirrors, is screened by a mile's thickness of rock, not only from the Sun's light but from everything else that comes from the sky, excepting only the neutrinos, ghostly atomic particles that can penetrate anything. 'Found any of them little strangers yet, Doc?' a tough-looking miner asks Ray Davis, as the graveyard shift ends. And Davis explains to his hosts that, just as a ton of their ore yields a fraction of an ounce of refined gold, so his huge tank can at best catch only a few neutrinos in a three-month run.

The first flight to the Moon, by Frank Borman and the crew of *Apollo 8*, at the end of 1968, was an epoch-making step in the exploration of the solar system of planets by new means. The samples of Moon rock, to be brought by the first astronauts who return safely from a landing on the Moon, will be of greater scientific interest than anything the *Apollo 8* astronauts could accomplish simply by taking their eyes into orbit round the Moon, where unmanned cameras had been before. Lunar rock samples will give information about the formation of the Moon and the history of the solar system; by contrast, the primeval, unmodified stuff of the Earth lies far below its weathered surface.

The astronomy of the planets has more in common with geology and meteorology than with the cosmic science of stars and galaxies. It is true that telescopes of all kinds also serve in planetary astronomy but the questions investigated are sufficiently different for the planets to merit separate treatment. Here I simply make a few remarks to put

*Homestake Mine. The biggest gold mine in the Western Hemisphere stands in the Black Hills of Dakota. In a gallery a mile below the surface is the oddest observatory in the world, designed for looking into the very heart of the Sun. Raymond Davis wears a miner's helmet in descending to his laboratory in the mine.*

*Surface of the Moon from a closely orbiting spacecraft,* Lunar Orbiter II, *showing part of the crater Copernicus. (NASA)*

them in their cosmic perspective. The Moon is the only astronomical object likely to be visited by man, for the next decade or two. The big novelty in planetary exploration will continue to be the unmanned rocket probes of ever-increasing sophistication. Already American *Mariner* spacecraft have returned, by radio, temperature measurements of Venus and television pictures of Mars. The Russians have parachuted a spacecraft through the atmosphere of Venus, though unfortunately it stopped transmitting before it reached the surface of that planet. Now we must look forward to spacecraft capable of landing instruments undamaged on the surfaces of Venus and Mars with the prime purpose, especially in the case of Mars, of looking for signs of life. Meanwhile, simpler systems will be hurled farther afield, to Jupiter and beyond.

Powerful radars on Earth provide another new technique for exploring the solar system. Echoes from the planets have cleared up long standing mysteries about how Mercury and Venus spin, have enabled surface features of Venus to be distinguished, through the mask of cloud that has always defeated optical astronomers, and have given highly accurate information about the positions and motions of the inner planets. Radio and infra-red emissions from Jupiter, the biggest planet of the solar system, are now studied with great interest. Jupiter is far from being a scaled-up version of the Earth; it would be nearer the mark to say it is a scaled down version of the Sun, a failed star not quite massive enough to ignite.

There will be plenty of fascination in the exploration of the solar system, justifying the heroic engineering necessary for shooting instruments across the interplanetary voids. By the end of the century men may be despatching a probe to Pluto and Uranus, at the bounds of the solar system. Yet even an effort of that magnitude will still be playing, so to speak, in the Sun's back yard. It is not that the solar system is small – rather, the universe is big.

The Moon is about 240,000 miles away, which means that radio signals from Earth to astronauts at the Moon take $1\frac{1}{3}$ seconds to reach them, travelling at the speed of light (186,000 miles a second). The Earth is $\frac{1}{23}$ light-second in diameter and the Sun is 8 light-minutes away. Pluto orbits out to a distance of $5\frac{1}{2}$ light-hours. The nearest star is more than 4 light-years from us, and the universe currently explored by the astronomers seems to be about 17 billion light-years in diameter, give or take a few billion. While the exploration of the Moon may tell us something about the events 4 billion years ago when

*The planet Jupiter. The largest of the Sun's family of planets has a conspicuous red spot; in this photograph by the 200-inch telescope one of the moons of Jupiter is also plainly visible, together with its shadow. (Mt Wilson & Palomar Observatories 200")*

the Earth, the Moon and the other planets were newborn, big telescopes observe events that occured long before the Sun came into existence.

### Which story?

Today's astronomers are men of the world, not like the distracted stargazers of the popular imagination. They argue robustly with governments about budgets. They adapt to new technologies in rocketry or electronics very readily. They are prominent among the scientific jet-set, flitting from continent to continent, for meetings, consultations or time on someone else's telescope. They are used to publicity because even if the man in the street's knowledge of astronomy is limited, his·interest is insatiable. The justification for spending public funds on fundamental science with no obvious utility is that new knowledge itself is of incalculable value. Unless the story is wrested out of the learned journals and shared as widely and rapidly as possible, how can we taxpayers tell what we are getting for our money?

The theorists outline the plot of the story, because that is their job – to interpret the broad significance of particular observations, and to pose new questions for their colleagues who man the telescopes. Progress in astronomy depends almost as much upon computers as upon telescopes; imaginative speculations on the blackboard are as necessary as new pictures of the real universe.

Pity the poor writer, though, who hears flatly contradictory state-

*The planet Pluto. The most distant of the Earth's planetary relatives is scarcely distinguishable from the background stars except by its motion, revealed here in photographs taken on two successive nights with the 200-inch telescope. (Mt Wilson & Palomar Observatories 200'')*

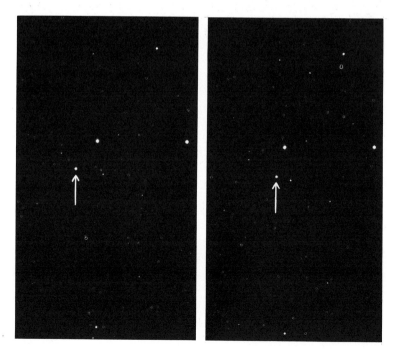

ments on succeeding days, from men of comparable distinction! That there is so much disagreement among theorists is a sign of how deep into preconceived ideas the new discoveries are cutting. It is also one of the most graphic illustrations one can wish for, of the human and subjective quality of science and of the part imagination plays in it. The picture of the scientist as a robot analyst was ever false. What distinguishes the scientist from other creative workers is the strict verification of bold ideas, to see whether they are right or wrong. To

*General plan of the universe.*

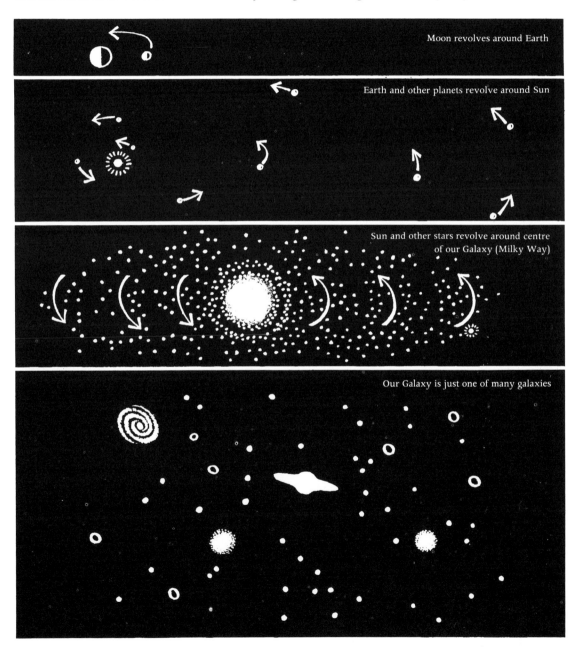

Moon revolves around Earth

Earth and other planets revolve around Sun

Sun and other stars revolve around centre of our Galaxy (Milky Way)

Our Galaxy is just one of many galaxies

navigate through this minefield of opinion in a foggy sea of fact, neither carelessly nor too cautiously, is the chief difficulty in publicising astronomy today. Nor is it comforting to know that developments in the early 1970s will show how good or bad one's final selection has been, of the ideas that tell the story.

At the observatories, too, one seeks opinions as well as facts. In practice there is no demarcation or closed shop separating the theorists and observers; the observational astronomer is himself a sophisticated physicist quite capable of making theories and posing questions of his own. Indeed he must be, because much of the art of being a good observer consists in knowing intuitively how to make the best use of priceless telescope time, and what the relation is between the power of the instruments and the problems to be investigated. Always, too many astronomers and questions besiege too few telescopes. At the observatory planning sessions – typically, once a month – nice judgement is needed between the pursuit of immediate excitements and those long-term investigations that provide the solid base of astronomical knowledge.

*Astronomer at work. Maarten Schmidt, the young Dutch astronomer who first recognised the extraordinary nature of the quasars, uses the 200-inch Palomar telescope to investigate them further. Only 16 astronomers from the whole world are privileged to work regularly with the world's biggest optical telescope.*

Every few weeks, the individual observer wins his few days or nights of telescope time. He packs his sweater, takes leave of his wife, and drives to the remote observatory. For this spell, even though surrounded by the electronic paraphernalia of modern astronomy, and with ultra-modern ideas in his head, he is not so different from the Megalithic priest of ancient Britain, who built his observatory of a stone circle and there communed with the mysterious sky.

The air is the enemy for astronomers. The thin envelope of gases and water enveloping the Earth cuts off most of the rays coming from the sky. If not, we should not be here, because much of the radiation of the Sun is deadly; until 420 million years ago, when a veil of ozone gas was drawn across the sky, screening most of the ultra-violet rays, life on this planet was confined to the relative safety of the oceans. Nevertheless, the atmosphere greatly limits the vision even of the most powerful telescopes. Turbulence in the air produces bad 'seeing', in the astronomers' phrase, which blurs the image in an optical telescope. A great deal of effort goes into the search for sites for new observatories, with this consideration in mind. The best 'seeing' at any working observatory is said to be that at Cerro Tololo, in Chile.

More obvious to the layman is the disruption of optical astronomy by clouds, and telescopes in dry climates certainly accomplish most work. It is not just a joke to say that one of the reasons why radio astronomy has been so strongly cultivated under the notoriously cloudy skies of England is that radio waves are not stopped by cloud. On the other hand, radio astronomy has flourished in comparable fashion in the sunnier climate of New South Wales. For the radio astronomer, thunderstorms may be a nuisance, but the real source of trouble for him is man-made radio interference.

Constant battles are fought to keep some radio frequencies 'quiet' for the sake of scientific investigation of the heavens, yet, when one of the New Jersey discoverers of the microwave background told me his work was frustrated by the radar at Kennedy Airport, it was not an isolated case. Communications satellites, high above the Earth, may also interfere with certain radio observations. Nor is it only radio transmitters that do so: electrical machinery and faulty electrical devices in cars, as everyone knows, can interfere with television reception, and radio astronomers suffer, too. When the Parkes radio telescope was pinpointing one of the earliest quasars, this historic experiment was safeguarded by university students stopping traffic for miles around. And when the hunt for pulsars began, the radio astronomers at Jodrell Bank found plenty – but they turned out to be electric fences on the local farms! Observatories are perforce lonely places. The growth of cities, with their electric lights, has dazzled many a famous optical observatory, including the Mount Wilson in California and Greenwich in England.

The blinding of astronomers by the natural atmosphere is more fundamental a problem. The air that looks so transparent to us is opaque to most of the radiation from the sky. Our eyes are adapted by the long biochemical experiments of evolution to respond to those wavelengths of light that pass unimpeded through the air; otherwise our eyes would be rather useless. Thus we have visible light which can be broken according to wavelengths, by a prism or fine grating, into the familiar rainbow spectrum: red, orange, yellow, green, blue and violet. To be precise the range of wavelengths reaching the ground is slightly wider than what we can see; it shades a little into the infra-red and the ultra-violet. The light of short wavelength – blue, violet and ultra-violet – is strongly scattered by the air, which is why the sky looks blue. The longer wavelengths, red and infra-red, are inherently more feeble forms of radiant energy.

Visible light, two thousand wavelengths of which span about a millimetre, is in any case only one form of electromagnetic waves, all of which travel at the speed of light and all of which come from objects in the universe. They extend over a vast invisible spectrum, on either side of the visible light in the wavelength scale, from radio waves measured in kilometres to X-rays and gamma-rays of wavelengths less than the diameter of atoms. Yet of all this great range of electromagnetic waves, only visible light and a narrow band of radio waves (from about one centimetre to ten metres) pass unimpeded through the atmosphere. All the rest are blotted out, except for a few special wavelengths in the infra-red – heat rays which happen to escape absorption in the air and can be picked up at ground with special detectors. In other words, 'messages' from distant stars and galaxies, forms of energy which could help to show the nature of those objects, are erased in the last few miles of their long journey.

From the time when Stone-age men began seriously to watch the slow drama of the sky, right down to the mid-20th Century, all we had by which to assess the universe was visible light. Telescopes and photographic plates augmented the power of the human eye, but even then the picture we had was a rather peculiar one. It depended as much on our own planetary biology as on the intrinsic nature of the universe. Think how odd your familiar surroundings would seem if you were not just 'colour-blind' but quite unable to see anything that was not actually coloured orange. The grass and trees, the sea, the blue of the sky – all these would be invisible to you. So it was with traditional astronomy, showing us serene majesty in the heavens, with the constant stars a silent solace for busy men.

### View from the top

Quite a different picture comes from radio astronomy, which makes use of the transparency of the atmosphere to short radio waves. The familiar lamps of the night sky, the stars, are simply undetected, even in big radio telescopes. The 'brightest' radio sources, excepting only the Sun, are objects invisible to the naked eye, and distant. One is the centre of our Galaxy, the Milky Way. Others turn out to be remnants

of stars that exploded long ago. And other bright radio sources are whole galaxies, very far away, which are exploding or at least in some remarkable distress. In other words, if we could look at the sky through radio spectacles we should see little of the familiar, tranquil universe. Our attention would fix on centres of upheaval and violent change.

Visible light emanates mainly from hot, dense matter, as in the stars. Radio waves come mainly from regions where diffuse, electrified matter travels at high speed through a magnetic field. So it is not really surprising that the old and new pictures are so strikingly different. But there are more pictures, too, which astronomers can see by rays in those other parts of the electromagnetic spectrum hitherto blotted from view by the air. This is, after all, the space age, and rockets and satellites burst right through the curtain of the atmosphere. They sometimes carry telescopes, with which new branches of astronomy are now being inaugurated: ultra-violet astronomy, which picks out particularly the hot, young stars; X-ray astronomy, which has disclosed a new type of star not easily explained; gamma-ray astronomy, which so far shows little except these energetic rays coming from the heart of the Milky Way. These discoveries have been made with unmanned rockets and satellites. The USA and USSR have plans for manned observatories in space, and some American astronomers are already training as astronauts.

Balloons carrying telescopes provide a more economical way of clearing at least the thicker layers of the atmosphere. And in the case

*Radio Astronomy Explorer.*
*This American satellite, launched in 1968, has a span of 1,500 feet with its aerials fully extended. Its purpose is to pick up cosmic radio noise of long wavelengths which do not penetrate the Earth's atmosphere. The Empire State Building is shown to scale.*

of another new branch of astronomy, looking at the sky in the infra-red (heat rays), the development of sensitive detectors has allowed important discoveries to be made even with ground-based telescopes – in dry climates. But one of the pioneers, the American physicist, Frank Low, prefers to fly his instruments in a jet plane to an altitude of ten miles, leaving far beneath them nearly all the moist air that interferes with his view of infra-red stars and galaxies.

The whole spectrum of electromagnetic rays from the universe has now been sampled, however cursorily, for the first time. Most of it has become accessible only in the 1960s and in these diverse new astronomies most of the work (and possibly the big discoveries) lie in the future. The most difficult wavelengths to work at remain those of a millimetre or just below, in the region of the spectrum where radio and infra-red merge.

The electromagnetic rays are not the only 'messages' from the universe that the atmosphere screens from us. Meteors, or shooting stars, are small grains of dust, hitting the Earth's atmosphere at great speed and burning up. Several thousand tons of matter adds itself to the Earth each year, from meteors. Occasionally big meteorites reach the ground without burning up completely. These can be examined and it turns out that they are as old as the Earth and similar in compo-sition. Spacecraft sample the finer dust that does not survive the collision with the air. Meteors, and also comets, are nowadays regarded as scrap material left over from the manufacture of the planets, some 4 billion years ago.

A quite different form of material also arrives continuously at the edge of the Earth's atmosphere, from the greater depths of space: the cosmic rays. These are swift atomic particles, sometimes possessing enormous energy. When they collide with atoms of the upper air they are greatly changed. Many transformations of matter and energy occur and what we register as cosmic rays at ground level bear little resemblance to the original radiation out in space. Again we are robbed of information they might give us about the universe. But here, too, detectors sent above the thick curtain of the Earth's atmosphere can catch that information before it is lost.

## Nature becomes self-conscious

Scientific research was always untidy, but never more than in our era when so many scientists are at work and the rate of discovery is so impressive. There is no master mind, nor master programme, imposing a coherent pattern on the activities of the astronomers. On the contrary, unexpected discoveries pop up from unexpected places, and the apparent importance of different lines of inquiry varies from year to year. Observatories co-operate impeccably and compete feverishly. In the disputes between rival theorists the objective wish to be correct is modified by the subjective wish to be right. The danger for the reporter, and his reader, is that the picture of current astronomy can fragment into uninstructive muddle.

One policy for avoiding such a hotch-potch would be to submit to one theorist or school of thought. But biased accounts may give false impressions of what is happening in the observatories and unless the theorist happens to be correct – a big unless' – the story may be a fairy tale, at least in parts. The greatest loss in such accounts is the picture of research in action – the clash of imaginations and the arbitration of facts – which is really more important for the enlightenment of mankind than oracular statements that this and this are so. Science is a process of discovery rather than a compendium of data. That we shall probably never know the whole 'truth' about the universe does not really matter very much; the fun comes in trying to find out.

The natural pattern of current astronomy, which I try to follow in this book, is provided by the cryptic unity of nature itself (belief in which is the chief act of faith of the scientist) and by questions astronomers ask today, as their predecessors have done for centuries:

Whence comes the fire of the Sun and the stars?

What is the explanation of unusual phenomena in the sky (in former days, comets and supernovae; nowadays quasars and pulsars)?

How is the universe constructed and what is our place in it?

How did the universe begin and what is its fate (and ours)?

If we were all-seeing supermen, these might appear naive and self-centred questions: there may be more significant ones which have not even occurred to us. But the human brain is itself a part of nature, fanned into existence by billions of years of sunshine acting on the molecules of the Earth. It is not perfectible in the immediate future, even if biologists should wish to alter the brain – which is a questionable ambition. What men make of·the universe at large is a product of what they can see of it and of their own human nature.

The visibility of the universe depends both on where we happen to be, in space and time, and on the instruments the astronomers have to hand: telescopes, photographic emulsions, computers, rockets and so on. Their human nature, as bipedal bundles of reason, imagination and prejudice, determines what questions astronomers ask, what sense they can make of the answers nature gives, and whether the human mind can ever grasp the full subtlety of the universe that gave it birth.

*Means of observation at different wavelengths and the typical objects that they show.*

← increasing wavelength

**Radio waves**　　　　　　　　　　　　**Infra-red light**

**Topside of the atmosphere**

**Ground level**

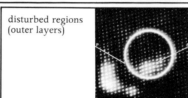

Means of observation

What the observations show

**Sun**

| disturbed regions (outer layers) | |
|---|---|

**Stars**

remnants of exploded stars (also pulsars, flare stars, embryonic stars)

embryonic stars (also cool stars)

**Our Galaxy**

spiral arms and other features

centre of Galaxy

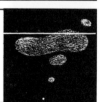

**Galaxies**

exploding galaxies and quasars

seyfert galaxies (also quasars)

**Universe**

microwave background

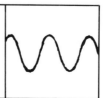

## Visible light

## Ultra-violet light

## X-rays and gamma-rays

| | | |
|---|---|---|
| sunspots and other events 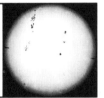 | solar flares  | disturbed regions  |
| stars in general  | young, hot stars  | X-ray stars (nature uncertain)  |
| obscured by clouds of gas  | | gamma-rays from Galaxy  |
| galaxies in general (also quasars and 'quiet' quasars)  | quasars  | exploding galaxy  |
| | UV background (helium) may exist  | X-ray background (hydrogen)  |

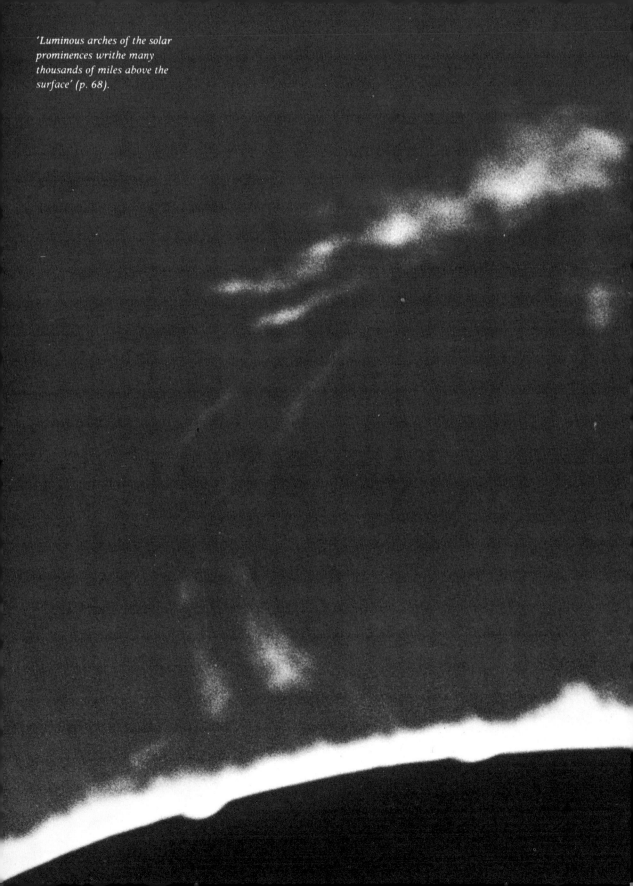

*'Luminous arches of the solar prominences writhe many thousands of miles above the surface' (p. 68).*

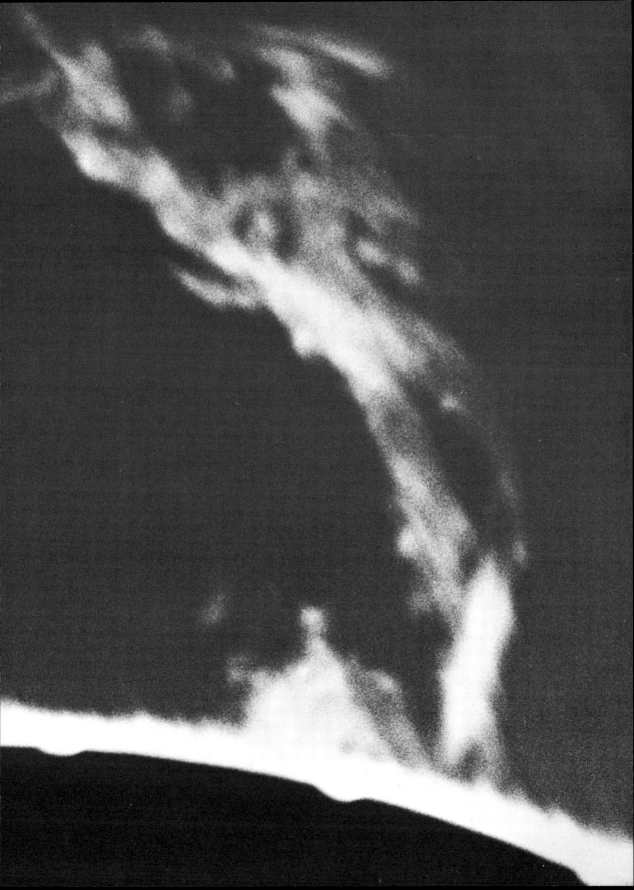

*Instant astronomy:* circa *1969*

Twinkle, twinkle little star
We know exactly what you are:
Nuclear furnace in the sky,
You'll burn to ashes by and by.

But tick, tick, tick pulsating star,
Now we wonder what *you* are:
Magneto-nucleo-gravity ball,
Making monkeys of us all!

And twinkle, twinkle quasi-star,
You're the limit, yes you are:
With such indecent energy,
Did God not say you couldn't be?

*Anon*

## A breath of violence

For twenty years, physicists from the University of Bristol have been flying packages of fine-grain film to great heights on balloons, to detect the 'primary' cosmic rays arriving from far away in space before they are degraded by the air. The cosmic ray particles burrow through the photographic emulsion, leaving distinct tracks. When the films are recovered and developed, the physicists look at the tracks under the microscope and deduce the properties of the particles. The typical cosmic-ray particle is the simple nucleus of a hydrogen atom, but Peter Fowler, at Bristol, studies the heavier chemical elements whose nuclei are also represented in the cosmic rays. He makes a remarkable discovery which bears closely on the origin and history of the cosmic rays.

In films recovered from a balloon flight in Texas made in collaboration with American scientists in September 1968, Fowler finds tracks corresponding to elements heavier than uranium. Uranium is the heaviest element found naturally on Earth, although nuclear physicists have made heavier ones artificially, using reactors and accelerators. All these very heavy elements are short-lived; they quickly change into uranium or still lighter elements, which is the reason why they do not exist on our planet. The simplest interpretation of Fowler's discovery is therefore that these heavy atomic nuclei have been made quite recently. Extraordinary circumstances are necessary for the new creation of heavy elements. The most likely origin for these cosmic rays is in exploding stars. They are a hot breath of violence from the depth of space.

*In Bristol University photographic film that has been flown at high altitude by balloon, the microscope reveals the tracks of cosmic rays, probably generated in exploding stars. The dense track, shown much enlarged, was made by a heavy element that no longer exists on the Earth's surface. (See also colour illustration facing p. 41)*

## Crab salad à la Chinoise

'Prostrating myself,' the chief computer of the calendar said to the Sung emperor of Khaifeng, 'I have observed the appearance of a guest star.' The less sophisticated Europeans did not notice the strange event in the constellation of Taurus, the bull, during that summer of AD 1054. Edward the Confessor was king of England; William of Normandy's countrymen were grabbing southern Italy, but another twelve years were to pass before he would change his designation from Bastard to Conqueror on the field of Hastings. The Moslems,

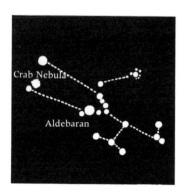

*Constellation of Taurus, showing the position of the Crab Nebula.*

despite their traditions in astronomy, were evidently too busy campaigning in West Africa to notice that the star they called Aldebaran had acquired a brighter neighbour. The records of the sharper-eyed Japanese and Chinese, re-examined in a striking joint study by modern historians and astronomers, provide one of the most useful accounts we have of an exploding star. In a matter of days it became as bright as the planet Venus and conspicuous even in daylight – 'reddish-white' in colour. Within a month it was fading.

Today the debris of that star is plainly visible in the big telescopes, as one of the most interesting objects in the sky – the Crab Nebula, a luminous shell of gas expanding in all directions at 700 miles a second. Red-glowing hairy filaments of hydrogen lace the white-glowing cloud. (See colour illustration facing page 40.)

Geoffrey Burbidge of the University of California, San Diego, remarks that nowadays astronomy can be divided into two sections: the astronomy of the Crab, and the astronomy of everything else. For many years the Crab has been a subject of intense study and theory-spinning, and yet it continues to heap surprise upon surprise, every bit as striking for contemporary astronomers as it was for Yang Wei-Te, the prostrate computer, 915 years ago.

When radar scientists, returning from the second World War, began radio astronomy in earnest, one of the very first cosmic radio sources was discovered by John Bolton, at Sydney in 1947. Taurus A, as it was called, turned out to be the Crab, broadcasting a protracted death-rattle of radio noise. In 1962-4, two groups of American scientist started not very optimistically, looking for X-rays from distant objects, using rocket-borne detectors fired above the atmosphere from the White Sands missile range. They were not disappointed. The first X-ray star to be pinpointed is in the constellation of Scorpius. The second in Taurus: again it is the Crab and the question arises, how is the X-ray energy produced?

The stuff of the Crab is riddled with magnetic fields and the nebula produces its radio noise, its light and possibly the X-rays too, by the effects of high-energy electric particles, electrons travelling through magnetic fields. The electrons are forced to spiral by the magnetism, and this makes them give off energy in the form of electromagnetic radiation. Relatively slow electrons emit radio waves, and lose energy only slowly: if they were started off at the time of the explosion seen in 1054 they would still be going strong. But white light is produced by the same sort of mechanism, and that means electrons more energetic to start with and more rapidly slowing down. In the course of a few hundred years, the Crab would have grown very faint, unless fresh supplies of electrons had come along.

*Opposite:*
*Crab Nebula – the debris of a star that was seen to explode in AD 1054. It is a strong source of radio waves and of X-rays. There is also a pulsar, or pulsating radio source, associated with the Crab. It is the visible star, arrowed here, which is actually flashing 30 times a second. (Mt Wilson & Palomar Observatories 200'')*

To make X-rays, very energetic electrons are required, and these lose their energy in the course of only a year. To find where the fresh

supplies of electrons can come from, Herbert Friedman, pioneer rocket astronomer at the Naval Research Laboratory in Washington DC, looks to the 'wisps of brightness' that surge out from the centre of the Crab every few months. Philip Morrison and L. Sartori have a different theory. They think the Crab emits X-rays, not because of the spiralling electrons, but simply because the cloud thrown out by the exploding star remains very, very hot – at billions of degrees. This particular argument will be settled by longer and more careful observation of the X-rays from the Crab.

Meanwhile the Crab comes up trumps once more. In 1968, radio astronomers from the University of Sydney join in the hunt for pulsating radio sources – the pulsars – first announced earlier in the year by Cambridge radio astronomers, as a new and baffling kind of object in the sky. Working at Molonglo with the big Mills Cross, an array of aerials in the form of a cross with arms a mile long, Michael Large makes a rich haul of additional pulsars. The first of them is one lying in a zone of the sky called Vela X which, like the Crab Nebula, is the remains of an exploded star. A possible link between pulsars and exploding stars has been suspected by some theorists, and here is the

*Part of the remains of an exploded star in the Southern Hemisphere. (Mt Stromlo Observatory)*

Green Bank. The 300-foot radio telescope, which first picked up the pulsar in the Crab Nebula, seen from below.

Arecibo. A radio astronomer searches the print-out of a computer for evidence of pulses from the Crab Nebula.

first evidence that it may be so – the first break in the otherwise bewildering accumulation of information about the ticking stars. But it may be just a coincidence in direction.

There are many remnants of stars which exploded long before records began. Only three of the really big explosions, called supernovae, have been observed in the past thousand years (in 1054, 1572 and 1604), but radio astronomers have found the characteristic shells of others. Michael Large and his colleagues at Molonglo promptly look at several other supernova remnants, but without finding another pulsar. That does not prove anything, either way, because pulsars are hard to detect even with the biggest radio telescopes. But another example of a pulsar associated with a supernova is certainly needed, to clinch the Australian discovery. At the end of 1968 colleagues at the radio observatories at Green Bank (West Virginia) and Arecibo (Puerto Rico) find another, NP 0532. It is a fast-beating pulsar and it lies in the direction of the Crab! Thomas Gold, of Cornell, by theoretical argument, traces the 'wisps of brightness' that keep the nebula alight back to the pulsar at its core.

For nearly a year, after the discovery of the pulsars, optical astronomers strove in vain to see a flashing light at the site of a pulsar as indicated by the radio signals. One night early in 1969, astronomers working at the Steward Observatory in Arizona spot the first visible pulsar, faint but flashing away at the same rate as the corresponding radio signals. Pulsars are visible after all, and the telegrams fly out spreading the news to the world's astronomers. 'It's easy to pick it up once you know where to look,' Donald Taylor of the University of Arizona tells me; he is one of the team making the discovery for which everyone has waited. Sure enough, confirmation comes quickly from the McDonald Observatory in Texas and from the Steward Observatory's neighbours on Kitt Peak, Arizona. The early observations show that the pulsar is very blue, and a strong emitter of ultra-violet as well as of visible light.

As for which pulsar is first to be seen, it is (need I say it?) NP 0532, the one in the Crab Nebula. It turns out to be a star well known in photographs of the supernova remnant, previously suspected of being the residual core of the star but not known to be flashing.

### Starr in Leo

The jazziest pulsar is CP 0950, in the constellation of Leo. When the radio astronomer feeds the signals to a loudspeaker rather than one of the more scientific recording devices, he can hear the various beats of the pulsars. The slower ones sound like heart beats, but CP 0950, at four beats a second, sets the feet tapping, and the illusion of a cosmic Ringo Starr is enhanced by the way the pulses vary in strength. But an astronomer's feet are likely to tap more in frustration, as he records these regular pulses coming from far away in space. What in heaven is a pulsar?

Amid a great deal of controversy about the possible nature of the pulsars, two deductions are scarcely disputed: that the pulsars are quite massive objects and that they are at the same time very small. The first conclusion follows from their amazingly good timekeeping qualities. They are far more reliable than most man-made clocks, despite the fact that they are throwing out vast amounts of energy many times a minute. They are losing speed only very slowly. They must have plenty of substance to draw on. The conclusion that the pulsars – or at least their radio transmitting regions – are also small comes from the sharpness of their pulses. Suppose a pulsar were as big as the Sun, and once a second it shook all over its surface and pumped radio noise into space. After a short interval, the radio noise from the nearest part of the Sun (the centre, as we see it) would arrive at the Earth. But the radio waves from the edges of the Sun, having an extra distance to travel, would arrive more than two seconds later. In other words, the signals from different parts of the Sun would smudge one another, and instead of sharp pulses there would be continuous rumble of radio noise. From the rate and clarity of the pulses, astronomers deduce that pulsars, or their active areas, must be a good deal smaller than the Earth.

Big mass and small size means the pulsars are very dense, with matter crushed forcibly together in a form we never encounter on Earth. Before the discovery of the pulsars, white dwarf stars were the most compact objects known. In such stars, at the senile stage with nearly all their fuel spent and no outward pressure of energy to support their own weight, the force of gravity compresses the matter in them until a thimbleful weighs a ton or more. The resulting hot stars are about the same size as the Earth, very much smaller than the ordinary run of stars, and much fainter, too. They were discovered in 1862 when investigations of irregularities in the position of the

*Anthony Hewish with Jocelyn Bell, the research student who drew his attention to the strange radio pattern that turned out to be the first pulsar.*

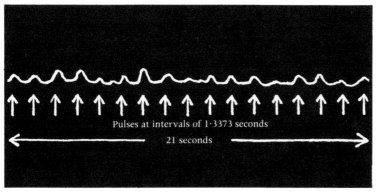

Pulses at intervals of 1·3373 seconds

21 seconds

*Pulsar pulses. The regular 'ticks' of the first pulsar to be discovered, CP 1919, as recorded at the Mullard Radio Astronomy Observatory at Cambridge.*

*Molonglo, New South Wales. The Mills Cross of the University of Sydney has aerial arms a mile long and has picked up more pulsars than any other instrument.*

bright star Sirius showed it to have a companion, an inconspicuous white dwarf.

But even a white dwarf seems too big for a pulsar, especially when new pulsars turn up with pulse rates of ten a second, or more. Indeed Anthony Hewish and his colleagues, who discovered the first pulsars, wondered from the outset if stars in an even more advanced state of collapse were involved – the 'neutron stars'. These are objects not observed before, but some theorists argue for their existence. A white dwarf is saved from even greater collapse by the resistance to pressure of the electrons it contains. But if the mass of the collapsing star is only a little greater than that of the Sun, its gravity will overwhelm the electrons, and collapse will continue until the whole star is like one huge atomic nucleus – a neutron star in which ten million tons of matter would not fill a thimble.

Opposing theorists, while not denying the possibility of collapse beyond the white-dwarf stage, question the existence of neutron stars. They doubt whether even nuclear forces can halt the collapse at the stage of a neutron star. An exhausted massive star, according to

them, just goes on shrinking and disappears from our universe. At the
moment, the neutron-star enthusiasts seem to be winning the present
round of argument; the pulsars may turn out to be somewhat different
from the theoretical notion of a neutron star, but the evidence says
they are something intermediate between a white dwarf and complete
annihilation by collapse. Both the high pulse rate and the location of
the pulsars near Vela X and the Crab Nebula give a great boost to the
neutron-star enthusiasts. The central debris from the explosion of
big stars is plausible stuff for making neutron stars.

The entrance qualification for the Pulsar Club is to own a very big
radio telescope, because the signals are really very faint – otherwise
they would have been noticed years ago. Some radio astronomers
exploit computers in hunting for new pulsars, using the machine to
search for snatches of rhythm in the radio records. Michael Large, the
ace discoverer of new pulsars, prefers to rely on his eyes, scanning the
tracks of the pen recorders.

## Gravity is greatest

However they work, pulsars illustrate, as do all other stars in their
various ways, the tricks that nature can play with the forces available
to it. When distinguished physicists meet to compare ideas at the
International Centre for Theoretical Physics in Trieste, Edwin Salpeter
of Cornell University seeks to explain the fascinations of astronomy
by pointing out that the long time-scales of the universe make rare
events visible to us. There are also, as he points out, great contrasts
– of high and low temperatures, of high and low densities of matter.

*Mills Cross. The underside of one
arm of the cross.*

Temperate conditions on Earth give us a very mild impression of
the forces of nature. Hurricanes and lightning strokes, earthquakes
and avalanches, the remorseless pull of the Earth's gravity and the
unforgiving inertia of a vehicle suddenly brought to rest – these are
the most unpleasant manifestations of natural forces, to which we
must now add the unnatural exploitation of another kind of natural
force, nuclear bombs. Yet, on Earth, hurtling masses are rather small,
gravity is not unduly oppressive, the chances of being hit by lightning
are insignificant and even the multi-megaton H-bomb is puny by
comparison with the nuclear reactions occurring in the smallest stars.
Among the solids, liquids and gases of our material world, we are
hard put to it to reproduce the common states of matter of the universe:
the diffuse near-vacuum of interstellar space, or the hot, electrified
gas (plasma) of the stars themselves. Our impression of the relative
importance of natural forces is also peculiar to our circumstances.

The three main kinds of forces known to scientists are nuclear,
electromagnetic and gravitational. In practice, physicists distinguish

between strong and weak nuclear forces, which are very different in kind, but the distinction is unnecessary for present purposes. If you had to invent a universe, you might well choose such a combination of forces, because they are complementary.

Nuclear forces are immensely strong, but operate over extremely short distances; they make possible those arrangements and rearrangements of the heavy constituents of matter that exhibit themselves in the nuclear furnaces of stars and in our own modest nuclear reactors. The 'bread-and-butter' nuclear reaction in the universe involves the fusion of four nuclei of hydrogen – the simplest and commonest raw material – to make helium. The resulting helium is not quite as heavy as four hydrogen nuclei. In other words, mass is destroyed and the star is slightly lighter. The missing matter is transformed into energy, at a very high rate of exchange (in accordance with $E = mc^2$, the famous law discovered by Albert Einstein, which involves the square of the speed of light, a very big number). The release of enormous amounts of energy in exchange for a little mass, destroyed by nuclear forces, is the secret of long life in the Sun and the stars. William Fowler of Caltech claims, with justification, that as much has been discovered about how stars work, by experiments with the accelerators of nuclear physicists, like himself, as by looking at stars with telescopes.

We ourselves live in an electromagnetic world. Everyday materials, including the tissues of the human body, are held together by the

*Accelerator at the California Institute of Technology. In the experimental area, beams of atomic nuclei from the Tandem Electrostatic Accelerator enable William Fowler and his colleagues to investigate nuclear reactions occurring in the extreme conditions prevailing at the heart of a star.*

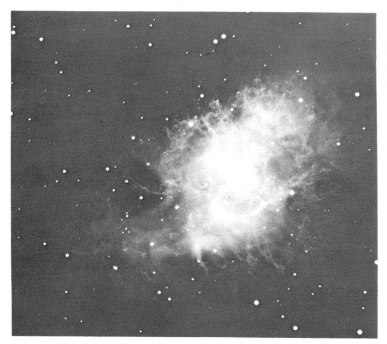

*Crab Nebula. In colour, the debris of the exploded star resolves itself into a white cloud and red-glowing filaments. The position of the flashing pulsar is shown on page 33. (Mt. Wilson and Palomar Observatories 200'')*

*Orion Nebula. Out of this bright cloud, a maternity ward of stars, new stars can be detected forming now. (Lick Observatory 120'')*

*Right:*
*Planet Earth, as seen by an American satellite high above South America. (NASA)*

*A normal galaxy, M 31, the nearby spiral galaxy in the constellation of Andromeda, is shown in all its glory in the 200-inch telescope. It is a huge assembly of stars like our own Milky Way.*

*An exploding galaxy, M82, in the constellation of the Plough. Faint streamers shoot out to both sides from the centre; the centre is also a source of radio noise. (M31 and M82 photographs, Mt Wilson and Palomar Observatories 48'')*

*Above:*
*Launching a huge balloon at dawn. It carried a Bristol University stack of film, which revealed heavy elements in the cosmic rays.*

*Right and below:*
*Michael Large, pulsar hunter, with the Mills Cross telescope at Molonglo. Bursts of pulses from the first pulsar linked to an exploded star are in the central tracks, below.*

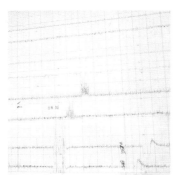

electromagnetic forces that operate between atoms. Rearrangements of atoms, in chemical reactions, liberate energy from food in our bodies and from coal and oil in our engines. The texture, colour and strength of familiar materials are all to be understood in terms of these electromagnetic forces, which are more obviously involved also in our electrical and electronic devices.

Gravity ties the Earth to the Sun and us to the Earth and stops us dropping head first into the abyss that starts at the top of our hair. The action of the Sun and the Moon on the oceans, to make the tides, is the only plain indication we have, on Earth, that there is more to gravity than dead weight falling to the ground. Although Isaac Newton taught us that every mass attracts every other mass with a gravitational force, the force is extremely weak between even the biggest of man-made objects. The rules for the prevention of collision at sea, for example, contain no mention of the gravitational attraction between supertankers.

The contrast between gravitational and nuclear forces is that one is strong where the other is weak – nuclear forces are ineffectual over long distances, while gravitation, so modest at short ranges, makes and breaks stars and moulds the entire universe. Gravity brings the matter of a star together and keeps it hot; indeed, it is only the outward pressure of electromagnetic radiation from nuclear reactions that prevents a star from condensing further and becoming ever hotter. We owe the very elements of which the Earth and ourselves are composed, to the power of gravity to turn on the heat necessary for their formation.

Electric currents and magnetic fields appear to play a relatively minor part in the drama of the universe, but this may be a mistaken impression that arises because astronomers are only just learning how to measure weak but extensive magnetic fields. As we shall see, magnetic fields produce striking effects at the surface of the Sun; they also figure in all the main sources of radio noise, from exploding stars and pulsars to exploding galaxies and quasars.

What we see and detect in the sky is to be interpreted, then, as manifestations of nuclear, electromagnetic and gravitational forces, and the greatest of these is gravity. Sometimes the forces collaborate harmoniously, sometimes an uneven conflict produces drastic results. And sometimes the resulting conditions are so extreme that astronomers lose confidence in the familiar laws of physics.

Neutron stars – if those are what pulsars are – have a very precarious stability, according to current gravitational theory. Only a narrow range of masses can reach this stage of collapse without overreaching it; that so violent an event as an exploding star should give birth to a neutron star is like using a bomb to make glass goblets. Yet pulsars exist, by the score, and the connection with exploding stars is hard to explain. Perhaps they gain stability from their magnetic fields or from fast rotation, opposing further collapse. Many theorists are unhappy about the neutron star – not least the nuclear physicists who are at a loss to say how so large a mass of nuclear matter should behave.

Thomas Gold of Cornell University is the man most confident that he understands the pulsars, and the discoveries linking them with exploded stars are a joy to him. Early in 1969, he sums up his ideas

*Gold's explanation of pulsars. The residue of an exploded star forms a compact, strongly magnetic mass, spinning at high speed. Matter oozing out is trapped by the field and whirled like a slingshot. It generates radio waves, light and cosmic rays as it breaks free.*

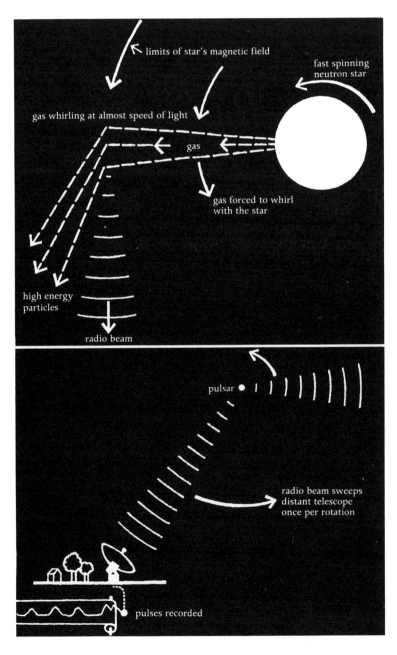

about pulsars: he seeks to explain, with a very simple mechanism, how the natural forces conspire to make the pulsar emit its pulses of radio waves and light, and also to power the mysteriously strong emissions from the expanding shell of the exploded star.

The core remaining from an exploded star, having used up its nuclear energy, has collapsed under gravity to make a neutron star a few miles in diameter. The gravitational energy so released goes chiefly into making the star spin at an enormous rate – perhaps a thousand revolutions a second to begin with. But the star also has an immensely compressed and powerful magnetic field, more than a million million times stronger than the Earth's magnetism. Any plasma – electrified gas – ejected from the surface of the neutron star is caught in the magnetic field and whirled at the same rate of rotation as the star. But when the plasma edges out to a certain distance from the star (30 miles, to start with) the whirling brings it almost to the speed of light. At that stage, the plasma not only generates light and radio waves but it breaks the bonds of the magnetic field, shooting off into space at enormous speed. Here, in Gold's view, is the source of energy that keeps the cloud of the Crab Nebula glowing. The radio and light are beamed by the whirling plasma so, if either the plasma emission or the magnetic field is lopsided, the waves are projected in our direction only once per revolution. As the neutron star loses energy, it slows down, to give the pulse rates of the pulsars so far discovered.

I summarise Gold's version of what pulsars are, in preference to dozens of other attempted descriptions, not because it can be regarded as proven but because it has the two virtues of being relatively simple and of not merely fitting but predicting the facts about pulsars, as they are known early in 1969.

### Juno's milk

Pulsars are not only a source of grief for astronomers who have to explain what they are; they are also a delight for those who wish to explore the architecture of the Milky Way. This is the lens-shaped collection of 100 billion stars or more, of which our Sun is a member. We see it edge on as a band of light across the sky. Our forefathers who, like present-day cosmologists, made up in imagination for what they lacked in knowledge, were pleased to regard it as milk spilled from the breast of a nursing goddess. It is dignified with the more learned-sounding name of Galaxy. By derivation the comparable aggregations of stars elsewhere in the universe are called galaxies, with a small g.

At Jodrell Bank, B. J. Rickett is using the well defined radio pulses from the pulsars, as if they were man-made measuring devices con-

*'Juno and the Infant Hercules' by Tintoretto, incorporating the legendary origin of the Milky Way.*

veniently placed far away among the stars, to measure the clouds of gas drifting in the disc of the Galaxy. His fellow radio astronomers have already mapped the general features of the Galaxy, but Rickett is interested in fine details in the clouds of diffuse gas that fill the spaces between the stars. His clue is the fluctuation in intensity of pulsar pulses over periods of several minutes. From careful measurements of the pulses from CP 0328, Rickett deduces that the fluctuations are due to electrons in clouds flicking across the radio path from the pulsar to the Earth. The clouds travel at speeds of ten miles a second or so, and are typically ten to a hundred million miles in diameter. They represent turbulence on a scale appropriate for the formation of stars.

Some of these clouds will one day condense, showing our Galaxy to be still fertile and capable of breeding new stars. Since the Galaxy itself condensed out of a turbulent gas cloud almost ten billion years ago, successive generations of stars have formed, including our own Sun.

The Sun is a star: a middle-sized, middle-aged citizen in the vast population that makes up the Milky Way. It is not 'merely' a star – there is nothing mere about it. Rather one should grasp that those pinpoints of light in the night sky are massive fiery objects like the Sun. To travel to the neighbourhood of another star is a feat that may always remain impossible for mortal men but let's go in imagination. If we look back the way we have come, it may be hard to pick out the Sun, to find our way home, because the Sun will not be particularly

*Northern stars, as seen from the neighbourhood of the Sun.*

*Northern stars, as seen from the neighbourhood of Alpha Centauri, a nearby star. The Sun becomes a member of the constellation of Cassiopeia. (After David Wallace, University of Maryland.)*

conspicuous. From Alpha Centauri, one of the nearest stars (actually, a triple star), the familiar W-shaped constellation of Cassiopeia is still evident in the northern sky, but it has acquired an extra member – that is the Sun. Other familiar constellations become distorted or broken up. What appear to us on Earth as groups of stars painted on the dome of the heavens are in reality objects at very various distances, so that perspectives change markedly as we make our imaginary interstellar flight. And all these conspicuous stars are relatively near neighbours, in the Galaxy.

If you should overhear an astronomer murmuring, 'Oh, be a fine girl, kiss me right now, sweetheart', jump to no hasty conclusion: he is simply using the standard mnemonic for the classification of stars by letters: O B A F G K M R N S, where O is the hottest and S the coolest. The Sun is a G star. Stars come in all sorts of sizes and conditions: big and small, young and old, bright and dim, and in a range of colours. There are many double stars revolving in company which seem to be an alternative to the arrangement we know best, in which our star has its litter of cool planets. Some other stars turn out to be double doubles or even triple doubles. Again, stars can be variable, repeatedly waxing and waning in brightness. Among the oddest stars are bright ones with very strong magnetic fields and with atmospheres full of rare elements, while others are intense sources of X-rays.

*Whereabouts are we, in the Galaxy?* For a long time, as a hangover from the days when first the Earth and then the Sun was supposed to be the centre of the universe, astronomers assumed that the Sun was right in the middle of the Galaxy. In the 1920s the Harvard astronomer, Harlow Shapley, rudely dislodged us and set us in a suburb of the Milky Way. He proved that the centre of the Galaxy was somewhere else, by showing that most of the globular clusters, collections of old stars that wheel around the centre of the Galaxy, lie to one side of us. It turns out that the hub lies in the direction of the constellation of Sagittarius, at a distance of 30,000 light-years. The outer edge of the Galaxy lies about 20,000 light-years away in the opposite direction.

The Sun and its planets lie neither exactly on, nor remarkably far from, the equator of the Galaxy, as defined by the ring of the Milky Way around us. We are a little to the north.

*What shape is the Galaxy?* Pictures of other galaxies, far away in space, show various possible shapes, from irregular blobs, through egg-shaped masses to neatly fashioned S-shaped or coiled spiral arms around a dense centre. The very appearance of the Milky Way tells us that the Galaxy is a flattened disc. But for optical astronomers like Harlow Shapley, life is made difficult by the dust that lies in the disc of the Galaxy and blots out their view of what goes on. Radio waves, on the other hand, travel relatively freely across this dusty disc.

*A globular cluster (NGC 6205). This dense collection of old stars is one of several such clusters that orbit around the centre of the Galaxy. (Royal Greenwich Observatory)*

*A galaxy similar to our own Galaxy, the Milky Way, seen edge on. The arrow shows the corresponding position of the Sun. This galaxy is NGC 4594. (Mt Wilson and Palomar Observatories 200'')*

What is more, the hydrogen gas that lies in the voids between the stars of the Galaxy is itself an emitter of radio waves, easily recognised because it has a characteristic wavelength of 21 centimetres. Dutch astronomers of the University of Leiden tuned into this hydrogen signal and began using it to map the shape of the Milky Way. A special trick was needed to distinguish different parts of the Galaxy lying in the same direction: it depends on the fact that everything in the Galaxy revolves slowly in orbit around the dense centre, but at varying speeds. As a result, the wavelength of the signal from the gas in different directions, and at different distances from the centre, varies slightly.

*Idealised plan of our Galaxy, the Milky Way, based on radio observations of hydrogen gas lying along the spiral arms. S marks the position of the Sun. Note the expanding cloud of gas emanating from the centre.*

*Opposite:*
*The dark clouds in our Galaxy obscure the optical astronomers' view, as in this nebulous cluster of stars in the constellation of Serpens. (Lick Observatory 120'')*

Other observatories joined in and the work continues, notably in the USA and Australia. As the great 210-foot Parkes dish sweeps the Milky Way, the radio signal from the gas between the stars abruptly increases and then decreases again, when the telescope encounters and leaves a dense region of the Galaxy. The combined results of work in the various countries have revealed convincingly the double spiral shape of our Galaxy. We can now state our whereabouts more precisely: we are in a spiral arm of the Galaxy that runs in the general direction of the constellation of Orion and is therefore known as the Orion Arm. One subtler point about the shape of the Galaxy is that it

is slightly deformed by the tidal effect of the large Magellanic Cloud, the nearest outlying galaxy.

*What is the history of the Galaxy?* In our own Galaxy, as in similar galaxies elsewhere, older (typically red) stars are concentrated in the hub of the Galaxy and in a great cloud of globular clusters and high-speed stars that surrounds the hub like a swarm of gnats. They do not fit into the main outline of the Galaxy and were obviously formed before it took its spiral form. The newer (typically blue) stars lie within the spiral arms of the disc. It is here that the gas – raw material for the stars – mostly lies; here we see new stars forming, and short-lived stars exploding.

The Galaxy has evolved. Subtle forces – gravity, magnetic forces, gas pressure – have evidently moulded a shapeless cloud of gas into the elegant spirals. The spiral galaxies – like our own – are as common as weeds in the universe, but astronomers are hard put to it explaining how the trick is done. The recipe for making spirals requires, for a start, neither too little gas (as in the elliptical galaxies) nor too much (as in the irregular galaxies). These other forms may represent other stages in the life of a galaxy, as it progressively converts its capital of gas into stars.

The Galaxy is held together by gravity: except at the very centre, it is ample to overwhelm any pressure of the gas to escape, and the stars and gas orbit round the centre, like the planets round the Sun, with the tips of the spiral arms trailing. But awkward questions arise. Why are there two spirals, and why do the arms not wind up, like a watch spring? Chia Lin of MIT has a theory which envisages waves of density sweeping through the disc and creating the spiral pattern. He does not invoke magnetism as a sculptural force, though it exists in the Galaxy.

In the closing months of 1968 two radio astronomers succeed where others have repeatedly encountered difficulty, in years of effort, in measuring the magnetic field in the Galaxy. Gerrit Verschuur at Green Bank and Graham Smith at Jodrell Bank use very different techniques. They also get very different answers – but then they are looking at different parts of the Galaxy.

Verschuur finds a part of the Milky Way – the so-called Perseus Arm – where the field is strong enough to cause a notable spreading in the wavelength of the 21-cm radio emission from hydrogen gas in the region. Verschuur reports a magnetic field strength of 20 micro-gauss. That is very weak compared with the magnetic field of the Earth but, operating over vast volumes of space, it represents a great deal of energy.

Pulsars are exploited by Graham Smith in his successful measurements of the magnetic fields in the Milky Way. Two characteristics of these pulsating radio sources make them very convenient for this

*Opposite:*
*Another galaxy like our own, seen from an oblique angle. The arms, rich with youngish stars, can be seen spiralling outwards from the bright centre. This galaxy is NGC 3031. (Mt Wilson and Palomar Observatories 200″)*

*Graham Smith, the radio astronomer who, at Jodrell Bank, has shown the pulsars to be useful aids in measuring the magnetism of the Galaxy's spiral arms. For many years he worked with Sir Martin Ryle at Cambridge.*

purpose. First, they lie close to the disc of the Galaxy. Secondly, they are themselves strongly magnetic, so that the radio waves they emit carry a special imprint of their magnetism. The waves are strongly polarised, which means they have been compelled to vibrate at a particular angle, when observed head-on. But, in travelling through the magnetic field of the Galaxy, this angle of polarisation rotates like the hand of a clock and long radio waves are much more affected than shorter waves. Smith therefore sets out to measure the magnetic field of the Galaxy by comparing the polarisation angles of different wavelengths coming from the same pulsar.

The first pulsar he looks at in this way, CP 0950, fails to reveal an intervening magnetic field, but the result, as Smith soon shows, is bad luck. In the direction of that pulsar there is no effective magnetic field, but the next two he looks at, CP 0328 and AP 2015, show clear rotations (corresponding to $3\frac{1}{2}$ and 2 microgauss respectively).

Although it may be unlikely that magnetism plays an important part in the overall shaping of the Galaxy, the very malleability of the fields as they are drawn out by the spiral arms may make them a kind of magnetic tape recording of the history of motions in the Galaxy. Moreover, on a smaller scale, particularly in gas clouds responsible for star formation, the magnetic field may be very influential.

*What is the future of the Galaxy?* The day will come when all the gas capable of giving birth to new stars will have been used up and the Galaxy will begin to fade in brightness, as first the most brilliant stars will burn out, and then the medium stars like the Sun. The typical stars of the spiral arms will be red rather than blue in colour, and the Milky Way (were we there to see it) would look more like blood. There are many currently negligible stars, much less massive than the Sun, that burn extremely slowly with a dull red glow – the so-called red dwarfs. They will burn for 100 billion years or more and, if the universe lasts that long, they will be the lowly remnants of the present magnificent Galaxy.

## Telescopes in space

The Orbiting Astronomical Observatory, or OAO for short, is a gang of robot astronomers, comprising the most elaborate unmanned satellite launched so far. It soared into orbit from Cape Kennedy on an Atlas-Centaur rocket on 7 December 1968. Four days later, when the satellite's basic functioning had been checked, six star trackers and eleven ultra-violet telescopes had to be switched on and tested one by one. Not until the New Year did the tests conclude and observations begin in earnest, after a period of anxious waiting for the American astronomers who have staked ten years of their careers on the idea of an OAO.

The nightmare possibility was that something might go wrong with this O A O as it did with its first and only predecessor, in 1965. On that occasion the experiments never began. An electrical design error caused a battery to overheat and, while engineers were trying to correct that fault by remote control, the observatory quietly expired. The long procedure of satellite design, construction and testing had to be endured all over again. $75 million of taxpayers' money was committed to the O A O. Should a similar wipeout occur again, on this second attempt, astronomy from space platforms might be discredited; in any case for the individual astronomers another chance would not recur for many years. If their experiments work as planned, on the other hand, they and the O A O stood to earn an honoured place in the history of astronomy, as a revolutionary development comparable with the introduction of the telescope and the radio telescope; it makes possible, for the first time, sustained and systematic observation of stars by radiation to which the Earth's atmosphere is opaque.

In the O A O control room at the Goddard Space Flight Center near Washington, engineers exchange signals with the two-ton satellite as it settles in its orbit 480 miles above the Earth. At Goddard and in two university cities, the astronomers wait for news. In the testing period, the fact that their spacecraft is the most complex automatic system ever put into orbit is not reassuring; a third of a million parts have gone into its construction and the failure of a single vital one could bring disaster for the project. But gradually the astronomers' anxieties fade, as the O A O passes all its tests.

*Orbiting Astronomical Observatory. This satellite, seen here under test at the Goddard Space Flight Center, was launched by the Americans at the end of 1968. It is gathering a rich harvest of information about ultra-violet emissions from young, hot stars.*

Fred Whipple and his group at the Smithsonian Astrophysical Observatory in Cambridge, Massachusetts, are out to survey the sky. Their system maps ultra-violet stars at a rate of 700 a day and measures their brightness at four different wavelengths, using four identical $12\frac{1}{2}$-inch telescopes fitted with TV cameras sensitive only to ultra-violet light. One day's operation with the OAO can give much more information about ultra-violet stars than has been acquired in 15 years of flights of ultra-violet telescopes in sounding rockets, which offered glimpses of only a few minutes each time. If there are surprising new ultra-violet objects to be found, the Smithsonian's search has a good chance of spotting them. In six months' operation it should survey about 50,000 stars covering a quarter of the whole sky.

Arthur Code of the University of Wisconsin, at Madison, leads the second team of astronomers working with the OAO. Their policy is to look more closely at relatively few objects – about 15 a day is the target. A battery of seven telescopes of various sizes will measure very carefully the intensity of each ultra-violet star or nebula at different wavelengths.

In order that the telescope will point at a selected region of the sky, the OAO has to be able to orientate itself. Six little auxiliary telescopes, or star trackers, serve this purpose. They steer themselves to remain fixed on selected bright stars in known positions in the sky. Using these reference points, the spacecraft as a whole can be turned in a chosen direction and held steady for long periods. Because its un-successful predecessor showed that the star trackers were not

*Astronomy by satellite. This early picture of a part of the sky, transmitted from the Orbiting Astronomical Observatory, shows several ultra-violet stars. (NASA)*

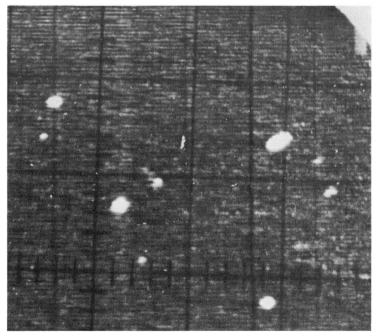

completely reliable, the new OAO also contains a system of gyro-scopes and acceleration detectors to help it tell which way up it is.

With the OAO, the lonely astronomer on the mountain-top is replaced by the machinery, computing and teamwork of an elaborate space flight. Each time the satellite passes over the eastern United States, the controllers milk it of its records of ultra-violet stars. An astronomer detached to Goddard from Whipple's group takes a quick look at the television pictures, to make sure they are showing something, but then the records are 'cleaned up' by computer before despatch to his colleagues in Cambridge. There, further computing is done, to characterise and catalogue the stars, while students mark their positions on a big wall map.

The scientific objectives of the OAO teams range from measurement of the amount of gas and dust in the gaps between stars to investigation of 'stellar winds' of stars like Gamma Velorum from which matter streams outwards at high rates. But the main pre-planned interest is in the very hot stars that shine particularly brightly in the ultra-violet. They are typically massive stars and, because of the rate at which they burn up their fuel, they must be short-lived. That means they must have been born much more recently than was the Sun, or most of the stars in the Galaxy. So the OAO is adding to our general knowledge of youthful, and potentially explosive stars.

## Turning gas into stars

Some maternity wards of stars are plainly visible in ground-based telescopes. Stars form, not singly, but in crowds, as a huge mother cloud of gas and dust gradually collapses under gravity and breaks into a litter of smaller, denser masses, which are embryonic stars. Just as the Crab Nebula is the most distinguished relic of an exploded star, so the Great Nebula in the 'sword' of Orion is the most studied region where new stars can be seen forming now. The Orion Nebula is a cloud of great beauty, as the colour illustration facing p. 40 shows.

Orion is studded with bright, young stars including one, FU Orionis, which 'switched on' – to the great excitement of astronomers – in 1936. They were really lucky to observe it, because the birthrate of such bright stars is probably only about one per 500 years, in the whole Galaxy. But as birth and infancy of stars takes a long time there are objects to be seen at different stages of the process. The Orion Nebula contains invisible objects which have been detected with the novel types of telescopes. These apparently coincide with pre-natal regions. At any rate strong infra-red rays come from these. So does a strange outpouring of 18-centimetre radio waves.

This 18-centimetre radio emission is characteristic of a gas con-

*The Cone Nebula. Another region where new stars are forming out of dark clouds. (Mt Wilson & Palomar Observatories 200'')*

*The constellation of Orion, showing the position of the Great Nebula.*

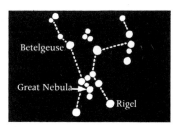

taining combinations of oxygen and hydrogen atoms (OH, incomplete water molecules). The mechanism that produces them is strangely like those recent human inventions, the maser and the laser, in which radiation energy is stored in a material and released by stimulation by other radiation. These OH regions in the sky are gas clouds, lit by very young, bright stars, but which are themselves forming further stars. OH is not the only chemical compound detectable by radio. Charles Townes, inventor of the maser, has turned to astronomy and, at the end of 1968, he and a group from the University of California, Berkeley, working with a 20-foot dish at Hat Creek report the detection of 1·25 centimetre radiation from ammonia in a cloud of gas and dust near the centre of the Milky Way. Later, water itself and other molecules become detectable.

The Japanese theorist Chushiro Hayashi, with his colleagues in Kyoto, has computed the pattern of collapse of gas clouds to form stars. It takes many thousands or millions of years for a cloud to

*Opposite:*
*The Orion Nebula. In this cloud astronomers have found radio and infra-red sources at the position arrowed. The emissions probably correspond to stars at a pre-natal stage. (Mt Wilson & Palomar Observatories 100'')*

George Herbig, authority on the youngest stars, picks out one of them in an eyepiece of the 120-inch Lick telescope, before breaking its light into its component colours and lines.

Star in the making. The luminous cloud is probably a precursor of star formation. (Lick Observatory 120")

form a condensation that ceases to be transparent: at that time the density is still low and the temperature is a mere ten degrees above absolute cold. Thereafter the condensation becomes much more rapid.

An impression of what the Sun was like at birth and in its infancy comes from the study of the new stars. George Herbig uses for this purpose the 120-inch optical telescope at Lick Observatory on Mount Hamilton. The stars he watches are T Tauri stars, with an unusually bright atmosphere, embedded in dust and also flashing in an irregular way – signs of 'growing pains' as the stars adjust their configuration in the search of stable maturity. He is able to estimate the ages of the stars from the amount of a rare element, lithium, they contain. Herbig is also finding, in the light of very young stars, signs that they are surrounded by complex material, of the kind from which planets may eventually be formed. Astronomers believe that the Sun's family of planets came into existence not long after the Sun itself was formed, and here is evidence that the process is repeatable for other stars. Herbig remarks: 'The planets are probably condensed out of material which was either spun off or blown off the Sun in its very earliest youth.'

Flare stars, which Sir Bernard Lovell picks up with the 250-foot radio telescope, may also be stars in raucous infancy, which have not yet found the stability of mature stars. Lovell is famous as founder and director of Jodrell Bank radio observatory in England, as con-ceiver of what, after eleven years, is still the world's biggest fully steerable radio telescope, and as an outspoken commentator on the space race. Time and again his big instrument has played a dramatic part in tracking American and Soviet experiments in space, and this is what has made headline news. But in fact the space-tracking operations account for only a very small fraction of the work at Jodrell Bank, and the observatory has taken a prominent part in the current exploration of the universe, including the approach to the discovery of the quasars, and the exploitation of the pulsars. Although Lovell himself has little time for his own research, he makes his mark in one line of investigation of which he says, 'This work wouldn't do for a young man who wants to get quick results.' His big telescope is working at the limits of its sensitivity, looking for rare events.

Ordinary stars pass quite undetected when even the biggest radio telescopes scan the sky. The radio emission is far too weak to be recorded. The Sun is an exception simply because it is so close. Until the discovery of the pulsars, Lovell's flare stars have been the only other stars to make any impression on the pen recorders that show radio noise arriving from the depths of space. He set out to pick up, by radio, stars that optical astronomers had seen occasionally flashing very brightly from time to time; the phenomenon looked like explo-sions on the surface of the star – like a king-size solar flare. Just four

*Hubble's variable nebula. The spluttering of this object is an accompaniment to star formation. (Lick Observatory 120″)*

*Sir Bernard Lovell. Known to the public chiefly as an outspoken commentator on the space race, he built the giant radio telescope at Jodrell Bank for astronomical purposes. He was personally responsible for the discovery of radio outbursts from 'flare stars'.*

hours after he began looking, Lovell registered a radio burst from a flare star but, as he says rather ruefully, it required five years' work after that, in collaboration with optical astronomers in several countries, to *prove* it. Now he has ample records, sometimes of quite astonishing explosions which nevertheless leave the stars intact.

## Seven ages of the Sun

In its relatively sedate middle age, the Sun's brilliant youth is far behind it, its bloated senility and shrivelled extinction far ahead. By looking at stars in different states, by laboratory research into nuclear processes, and by complex calculations by computer, astronomers can write the biographies of the Sun and other stars. The speed and course of events vary greatly according to the mass of the star.

The immense cloud of gas from which the Sun was formed possessed rich reserves of energy, especially of gravitational energy that could be released by collapse, and of nuclear energy in the form of hydrogen nuclei which could burn to make heavier elements. Also present, and important in its way, was the energy of the magnetic fields threading the cloud. The Sun, like the other stars, is a machine that converts the energy of the primeval gas cloud into heat and light. There is some

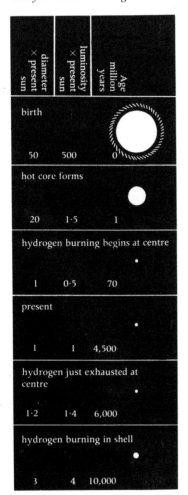

| diameter × present sun | luminosity × present sun | Age million years |
|---|---|---|
| birth | | |
| 50 | 500 | 0 |
| hot core forms | | |
| 20 | 1·5 | 1 |
| hydrogen burning begins at centre | | |
| 1 | 0·5 | 70 |
| present | | |
| 1 | 1 | 4,500 |
| hydrogen just exhausted at centre | | |
| 1·2 | 1·4 | 6,000 |
| hydrogen burning in shell | | |
| 3 | 4 | 10,000 |

uncertainty about details, and the more violent changes are hard to compute, but the theorists are fairly confident about the past and future of the Sun. It is a remarkable story of change in size and brilliance, as the Sun draws on various energy reserves in succession, until all are spent. In this account I use figures kindly collated by Ian Roxburgh of London University, an expert on the theory of stars.

1. First, the gargantuan infant shrank and flared in the mother cloud. When the Sun first came alight, it had not begun to burn its nuclear fuel. Its heat derived from gravity – from the falling together of the gas. But this initial capital of energy the Sun squandered in a mere eight million years, at first pouring out light energy 500 times faster than it does today, though from a dullish surface 50 times the diameter of the present Sun. The young star spun much faster than it does now, throwing off a substantial part of its matter. That process, and also the effect of the magnetism concentrated from the gas cloud, helped to moderate the rate of spin. The prodigal Sun went on contracting until it reached its present size, but was only half as bright.

2. At that point, the temperature at the very centre of the Sun became sufficiently hot to ignite the hydrogen – in other words to allow the release of nuclear energy to begin, in the conversion of hydrogen into helium. The Sun grew stronger again, until it reached its present brightness. More than four billion years later, it has scarcely changed. All the Sun's energy comes from a small zone in the very centre, where the nuclear reactions still proceed, using up hydrogen at a rate of 500 million tons a second. The inward pressure of gravity and the outward pressure of radiation from the centre are exactly balanced and the general structure of the Sun is very stable. Its diameter is a hundred times greater than the Earth's. During the next 1·5 billion years the Sun will become slightly bigger and brighter, but scarcely enough to disturb life on Earth.

3. Then something important will happen, though the effects will emerge only very slowly. The supply of hydrogen at the hot centre of the Sun will run out. There will still be plenty of fuel left in the surrounding mass of the Sun, but it is not transported to the central furnace. During four billion years following this first hint of old age, helium 'ash' will accumulate in the centre of the Sun, and hydrogen burning will continue in the zone immediately around this helium core. The rate of burning will actually increase, and the Sun will gradually enlarge until its diameter is about three times bigger, and the light output is four times greater. The climate of the Earth will already be greatly altered.

4. The next stage in the life of the Sun comes when the rate of burning in the shell around the helium core accelerates. Not only will

the Sun gradually increase its output of heat still further, and stifle the surface of the Earth, but the energy will lift the outer zones of the Sun against the force of gravity. After 600 million years of growth, the Sun will be fifty times its present diameter – a red giant star, pumping out energy 1500 times as intensely as the present Sun.

5. A brief but drastic change will then occur in the Sun. The helium ashes at the centre of the Sun will reach a temperature at which they too begin to burn, forming carbon and providing a fresh supply of nuclear energy. The onset of this process will be very sudden – the astronomers call it the helium flash – and when it happens the Sun will just as suddenly readjust its structure. In what may be only a matter of hours, the Sun will partially collapse again and greatly diminish its light output – *reculer pour mieux sauter*. The Sun resumes its giant growth after this setback; in the course of a further 30 million years, it will swell to a diameter 400 times greater than the present Sun. That means it will swallow, in turn, the planets Mercury, Venus, Earth and Mars. But when it becomes so bloated, it will be unstable, and it will puff off matter into the surrounding space, to form a planetary nebula expanding past the outer planets.

6. By this time, the nuclear energy available to the Sun will be all but exhausted. Apparently it cannot become hot enough to ignite the carbon in the core, which happens in more massive stars. As the fuel runs out, the pressure of radiation that props up the Sun against the force of its own gravity will be removed and, in the course of 50,000 years, the Sun will collapse upon its centre. This transition may be marked by short-lived explosions. At the end of it the Sun will retain only a hundredth of its present diameter. It will be a white dwarf about the size of the Earth, and the remains of the Earth and the other inner planets will be ground inside it. The collapse will release a fresh supply of gravitational energy, sufficient to ensure that the dwarf is very hot and bright. At first, it will be radiating energy at 50 times the rate of the present Sun. As it loses energy, it may freeze. As it changes from plasma into a solid crystal, it releases some additional heat.

7. With no fresh supplies of energy on which to draw, the dwarf Sun will steadily cool like an ember until, billions of years in the future, it will be first dull red, and then black. Sans heat or light, a small dead ball will wheel with other darkening ash-balls in a galactic graveyard of once-brilliant stars.

## Rotten at the core

A Canadian astronomer, Alan Batten of the Dominion Observatory, has discovered in 1968 the most massive star so far identified. It is 60

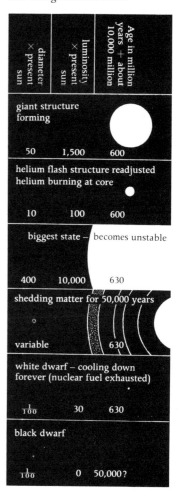

*Sun-old age and extinction*

| diameter × present sun | luminosity × present sun | Age in million years + about 10,000 million |
|---|---|---|
| giant structure forming | | |
| 50 | 1,500 | 600 |
| helium flash structure readjusted helium burning at core | | |
| 10 | 100 | 600 |
| biggest state – | becomes unstable | |
| 400 | 10,000 | 630 |
| shedding matter for 50,000 years | | |
| variable | | 630 |
| white dwarf – cooling down forever (nuclear fuel exhausted) | | |
| $\frac{1}{100}$ | 30 | 630 |
| black dwarf | | |
| $\frac{1}{100}$ | 0 | 50,000? |

*Life history of a massive star. Short-lived star (15 × mass Sun)*

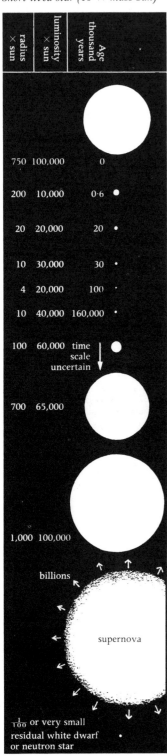

| radius × sun | luminosity × sun | Age thousand years |
| --- | --- | --- |
| 750 | 100,000 | 0 |
| 200 | 10,000 | 0·6 |
| 20 | 20,000 | 20 |
| 10 | 30,000 | 30 |
| 4 | 20,000 | 100 |
| 10 | 40,000 | 160,000 |
| 100 | 60,000 | time scale uncertain |
| 700 | 65,000 | |
| 1,000 | 100,000 | |
| | billions | |
| | supernova | |

$\frac{1}{100}$ or very small residual white dwarf or neutron star

times heavier than the Sun and it swings round with a companion only slightly less massive. Heavy stars are the young turks of the Galaxy: they shine brilliantly but burn out quickly. The dire effect of overweight on longevity is striking. If the Sun had retained just 25 per cent more matter, it would already have burned out. A star twice as massive as the Sun exhausts its nuclear fuel in one fiftieth of the active lifetime of the Sun, while a star such as Batten has discovered can survive for only a few million years, compared with the 10 billion years of the Sun. Having seen the forces of nature at work in the Sun, we can now look in more detail at the explosive fate of big stars.

Superficially, the first stages of life of a big star are like an accelerated version of the biography of the Sun, although the conditions in the very heart of the star are different: the centre is turbulent so that both the rate of burning and the volume first exhausted of hydrogen are greater than in the Sun. When the star grows to a giant, the pattern of events departs markedly from that in the less massive star. Genteel decline is ruled out when the augmented force of gravity ensures that the interior becomes much hotter. As a result, a series of nuclear reactions becomes possible, beyond the hydrogen burning and helium burning that occur during the life of the Sun. The carbon and oxygen formed by helium burning can themselves begin to burn as the temperature rises. In the centre of the star successive reactions continue to build up heavier elements and release energy, while the simpler forms of burning spread to the relatively cooler outer layers of the star. A star in this condition is a time-bomb, awaiting only the action of a detonator to make it explode, as a supernova.

The trigger for the supernova is, paradoxically, a loss of heat in the centre. By the time a lot of iron has accumulated in the core, the star is there running out of fuel because iron cannot burn. Making still heavier elements by nuclear reactions does not release heat, it consumes it, because iron is, in the nuclear sense, the stablest of the elements. The star would be doomed, even if other processes did not switch off its central heating.

Theorists argue about which is the more important, but the main cooling effects are:

1. the production of vast numbers of neutrinos, ghost-like atomic particles which can pour largely unhindered out of the core, taking energy with them; and

2. break-up of the iron, to make helium, abruptly reversing the laborious formation of the elements but also draining from the central region of the star the equivalent of the energy released during that formation.

The cooling makes the star rotten at the core. The central region collapses under gravity, in a matter of seconds. That process releases a fresh supply of energy which abruptly heats the outer regions of the star, detonates further nuclear reactions and builds up immense pressure. The explosion is monstrous, and overwhelms the force of gravity that holds the star together. New kinds of nuclear processes occur in the fireball of the supernova, expending energy in building up elements heavier than iron. Thus the greater mass of the star, transformed into a rich inventory of all the elements, is shattered and scattered into space, at hundreds of miles a second. These elements will be cannibalised in new generations of stars and planets.

There is a residue at the centre of the explosions. The core of the star possibly forms a white dwarf, like that predicted for the burnt-out Sun; more probably, in view of the current pulsar discoveries, it is a neutron star, strongly magnetic and fast-spinning.

Somehow, too, the exploding star creates cosmic rays – particles travelling at practically the speed of light. Theorists have been unable to tell how such immensely energetic particles could be produced in the instant of the explosion. Now Thomas Gold suggests that the cosmic rays are spun off continuously after the explosion, from the whirling neutron star.

*The Veil Nebula. The remains of an exploded star in the constellation of Cygnus. (Mt Wilson & Palomar Observatories 48″)*

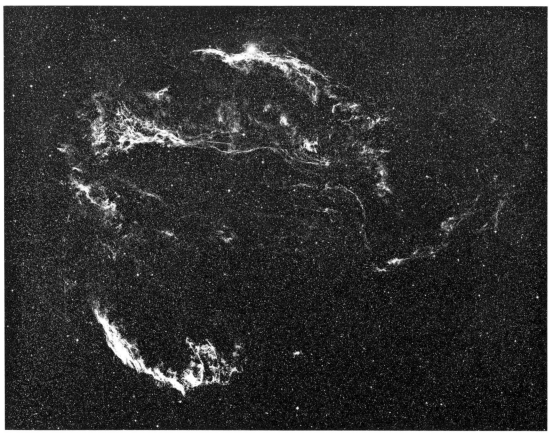

*A star explodes in another galaxy (1959). Left, the galaxy NGC 7331 appears normal. Right, and arrowed, a single star in the outskirts of the galaxy becomes a supernova, plainly visible across an immense gulf of space. (Lick Observatory 120'')*

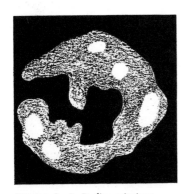

*Cassiopeia A. Radio emissions reveal the shell of an exploded star, the most powerful radio source in the sky. (After radio map from Mullard Radio Astronomy Observatory)*

Some of the stars we see burning placidly far away in the plane of the Milky Way must already have exploded, but the flashes of the explosions are still travelling towards us, and we have not seen them yet. It can happen any day or we may wait a hundred years for this exhibition of celestial fireworks. When the next supernova is seen from the Earth, astronomers will direct all the resources of modern astronomy on the event – the ground-based and rocket-borne telescopes, the spectroscopes for following the event in physical and chemical detail, and detectors to pick up the atomic radiation expected from the exploding star. The last supernova in our Galaxy was watched by Kepler in 1604, but that was a few years before Galileo introduced the telescope into astronomy. Indeed Galileo's interest in astronomical observations only began when he plotted the position of Kepler's star with a pen-knife fixed in his window in Padua.

While it will be a high day for astronomers, let's hope the next cosmic bomb does not go off too close to us. It could be calamitous for life on Earth, because the atomic radiation pouring into space from a nearby supernova would create conditions resembling all-out nuclear war. The astronomers could give some warning, because the flash of the explosion would reach us some months ahead of the dangerous particles. There would be no damage to cities, but we might have to evacuate them and go underground for a year or so. Survival would be economically difficult, even if protection from the worst effects of the radiation saved most people's lives. After the crisis was over, the cave-dwellers would emerge to find their fields in a crazy condition, from the combined effects of neglected growth and genetic mutation of plants. Most of the animals might be dead, with only the oceans providing some natural protection for its inmates. Equivalent events must have occurred dozens of times before in the course of the Earth's long history, and exploding stars may have precipitated some of the drastic biological changes known to have occurred during the evolution of life on this planet.

Astronomers do not have to wait for supernovae in our Galaxy to be able to study them. They occur in other galaxies and, because they are so bright, they are visible across immense distances. Fritz Zwicky of the Mount Wilson and Palomar Observatories began a systematic hunt for such events. Among the objects that Jesse Greenstein, of the same observatories, has newly found in his own study of supernovae in other galaxies, is one in the galaxy NGC 1058, an exploding star which shows enormous quantities of iron in the debris.

For astronomers today, the puzzle is not why stars explode, but why more of them do not explode. There are plenty of big bright stars around that seem strong candidates for supernovae, yet some mechanism saves them from that fate – or at least postpones it. To try to find out why such stars have not blown up is one of the tasks of the Orbiting Astronomical Observatory, that first satellite designed specifically for gazing at the stars.

## Neutrinos in them thar hills

Although their stories of the stars are impressive, the theorists are now waiting anxiously for news from the Homestake goldmine. They want direct evidence of the nuclear reactions they are convinced go on in the heart of the Sun, as a representative star. One man has the equipment for looking into the very centre of the Sun. Strange to say, his detector works as well at night as when the Sun is high in the sky. With his 100,000 gallons of perchloroethylene, a mile beneath the mining town of Lead, South Dakota, Raymond Davis of the Brookhaven National Laboratory runs a lab in a specially excavated cavern. Periodically he and his associates blow gas through the big tank of cleaning fluid, to churn it and draw out any radioactive atoms of argon that may have formed from the chlorine atoms in the tank. The experimenters spend a week doing it, four or five times a year. After elaborate refinement, the argon gas samples go for careful analysis, to show to what extent the atomic transmutation has occurred, during a three-month period.

The agents that can bring it about are neutrinos from the Sun. These have been mentioned before. Neutrinos are a very remarkable sort of atomic particles, produced in nuclear reactions. They have no electric charge and can travel unhindered through solid rock – right

*Tycho Brahe, Johannes Kepler, Galileo Galilei. By good luck, the three great founders of modern astronomy observed the most recent recorded supernova explosions of stars in our Galaxy. 'Tycho's star' blew up in 1572, and 'Kepler's star' in 1604. Galileo's first astronomical observations were made on the latter object. Although unconnected with their studies of the motions of planets, which paved the way for Newton, the 'new stars' helped to undermine official teaching about the perfection of the heavens.*

*Underground laboratory. Here Raymond Davis extracts from the fluid of his neutrino detector the few atoms of argon that register the arrival of neutrinos from the Sun.*

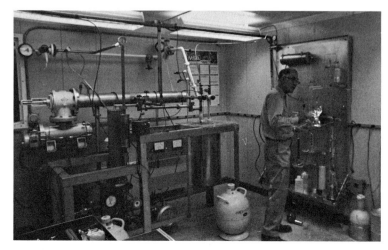

*Underground telescope. A huge tank of cleaning fluid, deep in the Homestake mine, serves to detect neutrinos, ghostly particles coming from the heart of the Sun.*

through the Earth, in fact. They can also come right out from the centre of the Sun in seconds instead of the thousands of years that other forms of energy spend in working their way to the solar surface. The very fact that neutrinos interact with practically nothing makes them hard to detect. Nevertheless, neutrinos of particular energy (to be precise, those produced by the breakdown of boron-8 nuclei in the Sun) have a small but definite chance of reacting with the chlorine atoms in Davis's tank to make the radioactive argon. It is because other, more reactive forms of radiation at the Earth's surface would completely mask this subtle effect of neutrinos that Davis puts a great shield of rock between himself and the surface.

Consternation came to the star experts in the summer of 1968, when Davis announced that there were far fewer neutrinos coming from the Sun than predicted. In the following months, while Davis made plans for a more sensitive detection system, the theorists began hastily to re-calculate their models. Their conclusion is that the basis of the theories can still be saved, provided Davis finds some neutrinos in 1969, with his improved equipment. Meanwhile they keep their fingers crossed, because the Case of the Missing Neutrinos is proving to be more of a challenge than they like. When the Homestake gold-miners heard that Davis was having difficulty in detecting neutrinos from the Sun, they tried to console him. 'Don't worry,' they said. 'We know it's been a very cloudy year.'

## Put it down to sunspots

Fortunately for us, the star to which our waggon, the Earth, is hitched is an undistinguished star and we have a chance to live. It does not mean that the Sun is a perfect, unblinking, unchanging object.

The nearby nuclear furnace in the sky provides the light and warmth that made life on Earth possible, and powers the machinery of green plants that keep life going today. Our planet itself, like the rest of the Sun's progeny, grew out of the dust and gas that was swirling round the Sun at the time of its formation. In a sense, we are still a part of the Sun, because its atmosphere has no sharp boundary. The space around the Earth is pervaded by a 'solar wind', though the bubble of the Earth's magnetism shields us from the atomic particles of which it is composed. For most practical purposes we are free to regard the Sun as it appears to the unaided eye – as a constant, bright and compact ball 93 million miles away. Yet in reality it is an explosively stormy object, in which fiercely hot material is thrown about by powerful magnetic fields, with repercussions on Earth and throughout the solar system.

The Sun's surface temperature of 5,800 degrees is maintained by rising bubbles of hot gas, which telescopes pick out as granules. Luminous arches of the solar 'prominences' writhe many thousands of miles above the surface and divide into comb-like streamers. And then there are the sunspots.

Astronomers in ancient China knew of the existence of dark patches on the Sun's disc. When they were rediscovered with the first telescopes, the sunspots added to Galileo's troubles with the Church, because they conflicted so thoroughly with the mediaeval belief in the purity of heavenly bodies. From records of sunspots over a couple of centuries, astronomers came to identify the 'sunspot cycle' of about eleven years. The frequency of sunspots rises and falls in a fairly predictable way, and so does the general storminess of the Sun.

Sunspots are the most obvious indication that the Sun is not the placid furnace it appears to be to the casual eye. They are the outward sign of upheavals on the face of the Sun associated with strong magnetic fields surfacing after lying low for hundreds of years in the interior. And, where the magnetic field builds up to very high intensity, great, localised explosions can occur – the solar flares. Solar flares occur near sunspots, brightening in a few minutes and more slowly

*A solar prominence. A huge luminous cloud follows a magnetic loop 200,000 miles above the Sun's surface. (Mt Wilson and Palomar Observatories)*

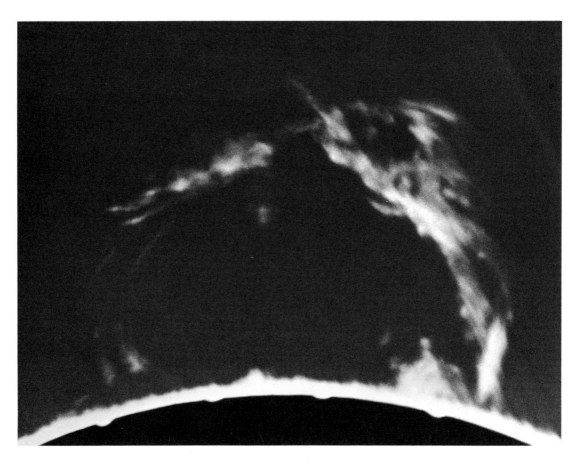

*The Sun's corona. The hot atmosphere of the Sun becomes plainly visible at times of eclipse. Filaments from the poles give an impression of the magnetic field. (Royal Greenwich Observatory)*

*Sunspots. Cool, dark regions are the most conspicuous sign of magnetic upheavals near the visible surface of the Sun. In this photograph, taken at the start of a period of high activity on the Sun, the spots lie towards one pole. (Mt Wilson and Palomar)*

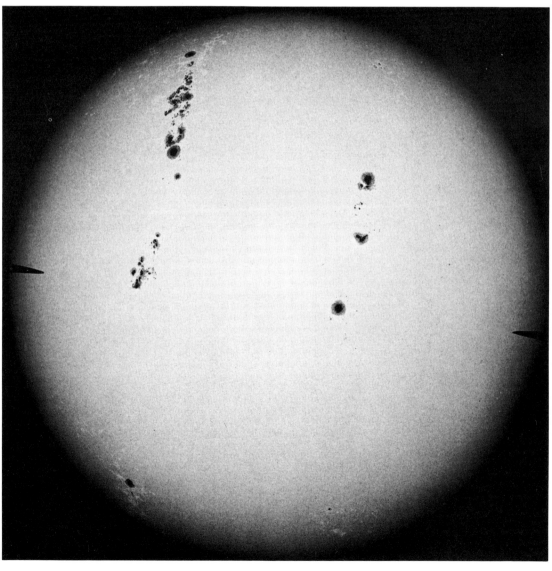

fading. Related disturbances produce strong X-ray emissions, detectable with rockets and satellites.

Connections between events on the Sun and phenomena in the Earth's environment have been dominant themes of the great international exercises of the International Geophysical Year (1957-8) and the International Years of the Quiet Sun (1964-5) and the current Solar-Terrestrial Physics programme. Many of those esoteric-sounding experiments included in scientific satellites and space probes are really concerned with solar-system weather. The confirmation of the existence of the solar wind by Russian and American space probes must be counted as one of the first great discoveries of the space age, along with the detection of the Van Allen belts of radiation, and of the X-ray stars.

The solar wind, a continuous breeze of electrified gas, blows at several hundred miles a second outwards from the Sun, in all directions. Here at last is the true explanation of why tails of comets always point away from the Sun – they are like smoke in the wind. The solar wind itself both entrains and responds to magnetic fields in tortuous ways, and there are magnetic barriers to its escape from the Sun. The wind may emerge in fountains called 'spicules'; in a period of 10 minutes a spicule surges to a height of about 5000 miles above the visible surface of the Sun and falls again. At each moment there may be 100,000 spicules dotted about on the Sun.

When sunspots are in evidence, the solar wind becomes gusty, and it is capable of disturbing the Earth's magnetism, in creating the bright draperies of the aurorae as atomic particles slam into the upper air over the poles, and modifying the electrified layers of the ionosphere that make normal radio communication possible.

Magnetism leaves its imprint on the light from the Sun. Half a century ago an American astronomer, George Hale, discovered the trick for measuring solar magnetism from a distance of 93 million miles. He was the first to show that the Sun had any magnetism at all, and he did so by spotting the small but significant smudging of the wavelengths of light emitted by particular atoms of the Sun. In the late 1950s, using a similar principle, Horace Babcock of Mount Wilson and Palomar Observatories found that the Sun swaps its magnetic poles from north to south every eleven years, in time with the sunspot cycle. In the latest instruments the technique has been so refined that it is possible to watch the minute-by-minute changes in magnetic fields across the face of the Sun.

With the great 60-inch solar telescope at Kitt Peak, astronomers make pictures of the Sun's magnetism using a big spectroscope that splits light finely into its component wavelengths. William Livingston causes the image of the Sun to scan over the slit of a detector that measures the magnetism at each point; from this information he builds

up charts that reveal the spotty magnetic zones of the stormy Sun and, in particular, regions of 'north' and 'south' magnetic poles lying close together. Between them, strong fields arch through the Sun's atmosphere, creating a jungle of magnetic forces on the disturbed face of the Sun. With the same telescope, Neil Sheeley takes snapshots of the magnetism at different levels in atmosphere, by photographing only the light of a particular element, as altered by its magnetic environment. The 'spectroheliographs' so made show that the patterns of magnetism near the surface are blurred in the higher levels of the Sun's atmosphere.

The Kitt Peak astronomers have made a new and perplexing discovery: that the lower levels of the Sun's atmosphere rotate faster than the visible surface between them. Patterns of magnetic activity travel across the disc, as the Sun spins on its axis at a rate of about one revolution a month. Being fluid, the Sun spins faster at its poles than at its equator. Theorists have also deduced, from the existence of flattening at the poles, that the core of the Sun is rotating very much faster than the surface. This gives the reasonable picture of the fast-spinning core of the Sun dragging the outer layers around with it. But the Kitt Peak observations, if correct, tell a contradictory story: that the atmosphere is pulling the visible Sun around. 'Instead of the

*Exploding Stars*

*Kitt Peak solar telescope. A flat mirror on the top of a tower reflects sunlight down a long tunnel to a 60-inch focusing mirror whence it is returned to the observing room. (See also colour illustration between pp. 136-7.)*

rotation of the very massive Sun being governed by its interior, as you would think,' Livingston says tentatively, 'it's actually being controlled or driven by the very tenuous gases which extend through the planetary system.'

*Culgoora, New South Wales. Part of the ring of 96 dishes of the huge radio telescope specially designed for watching events on the Sun.*

Strong magnetic fields and swift charged particles are also a recipe for producing strong radio emissions. That the Sun emits radio noise was discovered in 1942 by a British army physicist, Stanley Hey, when he was investigating 'jamming' encountered by radar stations. Although it subsequently turned out that the Sun gives off radio waves even when it is clear of spots, radio astronomers have found by far their greatest interest in radio signals associated with storms on the Sun. When a gust of solar wind travels outwards from the Sun at a steady speed, it triggers radio emissions of increasing wavelength, from each layer of the Sun's atmosphere in turn.

The biggest radio telescope designed particularly for looking at the Sun is at Culgoora in New South Wales. This 'radio-heliograph' produces motion pictures of the Sun's weather. The main instrument consists of 96 dishes arranged round a circle six miles in circumference. Each dish is 45 feet in diameter and automatically steers itself to follow the Sun. Two hundred miles of wire carry the signals from the dishes to the centre, where they are blended to give a sharp picture of one point on the face of the Sun. The system scans the radio Sun once every second and reveals the effects of magnetic activity in a continuous sequence of television-like pictures.

Paul Wild, the radio astronomer who conceived this unique tele-

*The Sun by radio. This 'radioheliogram', obtained with the Culgoora telescope, shows the eruption of a prominence (P) triggered by a flare (F) which occurred at a distant part of the Sun a few minutes earlier. The white circle markes the visible disc of the Sun; the radio emissions come from the surrounding atmosphere. (CSIRO)*

scope, is well satisfied with the records of very striking events it is producing. Most dramatic are those showing how a shock wave from one storm centre can skim across the Sun and trigger another event a million miles away, a few minutes later; or how a collision of criss-cross magnetic fields can spark off explosions at several places on the Sun, simultaneously.

The physicists who try to give their fellow men the benefit of controlled nuclear fusion as a source of energy have to create very hot, electrified gas, or plasma – to put, as it were, the fire of the Sun in a bottle. These experimenters lack the advantage of the Sun's gravity to hold their hot plasma together, and they have spent many years trying to pin it down with strong magnetic fields, long enough for the nuclear reactions to occur. But the plasma shows an exasperating talent for wriggling out, expanding and stopping the wished-for reaction. The same sort of vicious competition between plasma and magnetism occurs on the surface of the Sun.

The plasma near the Sun's surface becomes entangled in the local magnetic fields like balloons in a net. The magnetic field tends to hold down the plasma, or let it merely billow, as in an arching prominence. But at the same time, the plasma tends to escape, dragging the magnetic field with it. The struggle rages in centres of activity, each covering about a tenth of the Sun's visible disc and persisting for weeks or even months.

Sunspots are regions of strong magnetism where normal circulation is impeded, so that the plasma becomes cooler and relatively dark. The same magnetism prevents the local jets of plasma being injected into the solar wind. Plasma therefore accumulates, pressing ever more strongly against the magnetic 'net'. After a while, a long streamer is to be seen stretching outwards from the Sun for a distance of more than twice the diameter of the Sun. According to Peter Sturrock of Stanford University, it is a package of pent-up plasma, stretching the Sun's magnetic field at the climax of the struggle. Something has to give, and the magnetic field tears, at the base of the streamer. The release of the tension drives part of the plasma hard down on the Sun's surface, where its impact creates the visible flare and X-ray emissions. The rest of the plasma cloud lurches out into space, spreading as it goes, but guided by the general magnetic field of the Sun in a gentle arc towards the planets.

Can these explosions on the Sun be predicted, say, a week ahead – especially for the sake of astronauts and supersonic air passengers, who may be grounded because of the risk of high-energy atomic radiation thrown out by major flares? A leading Czech astronomer, Zdenek Svestka of Ondrejov Observatory, takes a pessimistic view. He recalls that an active region appeared on the Sun on 3 July 1966 and two days later its structure indicated that it might develop a

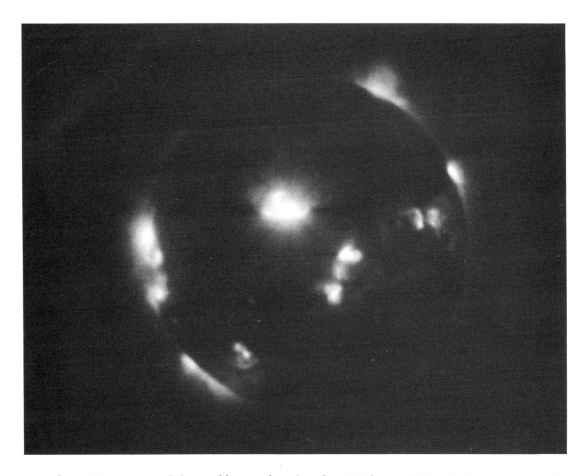

major flare. Scientists round the world were alerted, and on 7 July a major flare duly occurred. Yet, Svestka insists, 'Nobody could have the slightest idea, five days before the event, that a major flare might appear in that region.'

The Space Disturbances Laboratory at Boulder, Colorado, makes daily forecasts of the likelihood of major flares on the Sun, for one, two or three days ahead. The scientists examine the Sun, region by region, looking for active sunspots, magnetic complexity or strong radio emission at millimetre wavelengths. Their results are much better than they would get by guesswork, but still far from perfect. Forecasting solar weather is still like forecasting English weather – the weathermen need a certain amount of luck, as well as scientific skill.

*The Sun by X-rays. A remarkable photograph obtained by R. Giacconi and his colleagues with a rocket-borne telescope in 1968. (American Science and Engineering, Inc.)*

### Eggheads of the Galaxy, unite!

A 'wiggler' is a star that is neither completely steady in its motion in the sky nor plainly orbiting around a companion in a double star. If, in the very careful measurements of the astronomers, a star shows

*A chemist investigates the origin of life. At the Salk Institute in California, Leslie Orgel from Britain puts in a flask the gases thought to have existed in the atmosphere of the early Earth. Sparks in the gas, equivalent to lightning flashes, create chemicals which dissolve in the water, representing the seas. The chemical 'soup' so formed is appropriate for the emergence of life.*

a slight wiggle, but not large enough to be accounted for by a dark star, the most probable explanation is that it possesses a family of planets. like the Sun's. Although convenience makes us think of ourselves going round a steady Sun, in fact our own star wiggles off course as the heavy planets wheel around it. Such movement is perceptible only in nearby stars, at least until large telescopes are put into orbit above the Earth's atmosphere.

Wiggling, and the signs of complex material in the neighbourhood of newborn stars, are the only direct evidence we have about the possible existence of planets attending other stars. So far there is only one really good example of a wiggler produced by planets of mass comparable with our own solar system. The ancient question of whether we are unique in the universe is more nagging than ever, just now. The discovery of how immense the universe really is leaves us with a sense of loneliness like travellers in a wide desert.

In more practical terms, we can see how easy it would be for civilisations on distant planets to communicate with one another by radio. Already American radio astronomers have looked seriously, though in vain, for radio signals from the region of two promising nearby stars, and the Russians are beginning a more extensive programme. So prevalent and serious is the belief, among present-day astronomers, in the possibility of interstellar communication by radio that, when Anthony Hewish and his colleagues first detected the steady radio beat of a pulsar, they labelled their records 'LGM' – for Little Green Men.

The current theories of how the Earth and the other planets formed, from gas and dust eddying round the young Sun, make it seem a likely thing to happen to other stars. Planets should therefore be quite common in the universe. Not every planet will do, of course; of the Sun's family, only the Earth looks capable of supporting life as we know it – although, of course, great interest attaches to the plans for sending automatic laboratories to the surface of Venus and Mars, to look for primitive organisms or the chemical precursors of life. Among chemists and biologists, there are plausible hypotheses, backed up by laboratory experiments, telling how the Sun's rays brought living organisms into existence from a chemical soup on the early Earth. Few scientists now believe in a special act of creation on this planet, that could not be repeated many times elsewhere.

Then one has to start multiplying the odds against the formation of a suitable planet of a stable star, against the origin of life, against the emergence of intelligent beings surviving for a long period, and against them evolving both the technology and the interest for communicating with other civilisations. The combined odds against all this may be very long indeed, and yet the number of stars, even in our own Galaxy, is so very large that it is still difficult to believe we are unique.

Human ability to communicate across great distances being so recently acquired, we must assume that 'the others' have been at the game much longer, so it will be up to us to adopt the mode of communication they have chosen. There will be great cultural obstacles to conversing with beings differing widely from us not only in knowledge but in form and habits, too. Nevertheless the laws of mathematics have an abstract quality that circumstances cannot alter, so they provide a basis for devising codes and languages for communication. Lest 'conversation' should be taken too literally, remember the vast distances between the stars, and the fact that even if we do locate a communicative civilisation at a relatively near star, messages may take ten or a hundred years to pass between us. We may find ourselves posing questions and reminding our grandchildren to tune in for the answers.

Without exaggeration, making radio contact with eggheads on other planets could be the most important event in human history, and there is no reason to say it is impossible within the next few decades. Whether it will be a happy event is another question. We may find ourselves conversing with angels or devils, who teach us great wisdom or seek to brainwash us from afar. One distinguished American biologist, George Wald, admits the likelihood of contact to be quite considerable but his motto is 'Don't answer!' His chief worry is the discouragement and loss of dignity which we may suffer in learning from a superior technological civilisation, and which may make human enterprise seem pointless. Perhaps Wald is right. What satisfaction would there be in astronomy, for example, if somebody told us exactly what the pulsars and quasars were without leaving us time to find out for ourselves?

NGC 1275 – 'a quasar in slow
motion' (p. 97).

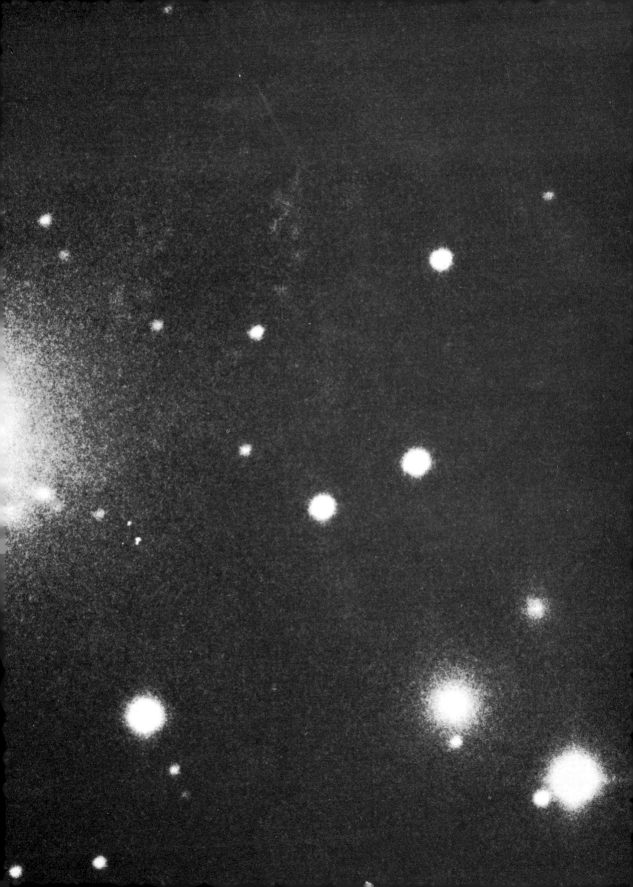

*Farther shores of time*

The southern sky is more beautiful than that of the northern homes
of most astronomers, and the unfamiliarity of the southern constellations forces the visitor to look at the stars afresh. The central regions
of the Milky Way adorn most prominently this other 'half' of the
universe. The world is purblind in one eye, because the big telescopes
are concentrated in the northern hemisphere and a substantial part
of the sky is invisible to them. The discoveries of the Australian radio
astronomers show how disabling is the lack of any optical telescopes
south of the Equator greater than 74 inches. The situation is changing
with the construction of telescopes of twice that aperture, in South
America and Australia.

Among the special objects that the southern polar skies offer the
astronomer are the clouds observed by Magellan the navigator: two
faint but quite large nebulous masses lying between the Milky Way
and the triangle of Phoenix. They look as if they are parts of our Galaxy
that have broken off, nor is the eye very misleading on this score.
The Magellanic Clouds are, in fact, other galaxies, the closest neighbours of the Milky Way. They lie nearly 200,000 light-years away – an
unremarkable distance in galactic terms, being only two or three dia-
meters of the Milky Way.

The most conspicuous galaxy in the northern hemisphere is fifteen
times as far as the Magellanic Clouds. It is barely visible to the unaided
eye, as a smudge of light sitting in the lap of Andromeda. Unlike the
ragged Magellanic Clouds, the Great Andromeda Nebula, or M31, is
an elegant, flattened, spiral galaxy; we happen to see it from an angle
that makes it look oval, as in the colour illustration between pages
40–1. Though like the Milky Way, it is bigger. The light that falls
from it into the Earth's telescopes left M31 nearly $2\frac{1}{2}$ million years
ago. At that time, our ape-like ancestors, chasing game in Africa with
crude weapons, were beginning to exhibit a few human qualities, but
*Homo sapiens* had not been invented.

Such a gloss cannot fully convey the remoteness of M31, nor can
the bald statement that it is 25 billion billion miles away. In any
case, astronomers encounter objects several thousand times farther
than M31, from which the radiation now detected was launched so
long ago that we see these objects as they were before life began on

*Opposite:*
*Large Magellanic Cloud. This*
*irregular mass of stars, conspicuous*
*in the southern sky, is the nearest*
*galaxy to our own. (Lick*
*Observatory 6")*

Earth, or before the Sun was born with its sibling stars from the mother gas. To these farther shores of time the distances are reckoned in billions of light-years. The human capacity for mapping an ocean on the back of an envelope nevertheless permits us to visualise, by a mere transformation of scale, the universe of galaxies.

Picture, then, the sphere of space visible to the most powerful telescopes. Put our own Galaxy, the Milky Way, in the centre, not because it is the middle of anything but because it happens to be our vantage point. Very close to it put the galaxy M31 and, in the little bulge of space so defined, pack 15 assorted galaxies, mostly small but including another spiral (M33) as well as the Magellanic Clouds. This is a local group of galaxies. On the opposite side, but still close to the centre, and splayed across 30 degrees of the sky as seen from the Milky Way, put the Virgo cluster. This is a much more populous cluster of galaxies, numbering at least several hundred. Then fill the large remainder of the visible sphere with other galactic clusters, judiciously spaced, some big like the Virgo cluster, some small like our own. Each galaxy, and cluster, represents a huge cargo of matter battened down by gravity.

If, working outwards from the centre, you begin to weary when you have placed some billions of galaxies, you may take comfort that at this stage the great optical telescopes, too, begin to flag, and populations at the boundary of visibility become doubtful. At this range

*A cluster of galaxies, showing many different types. (Mt Wilson & Palomar Observatories 200″)*

ordinary galaxies, even big ones glowing by the light of a hundred billion stars, make no detectable impression on the sensitive photographic plates of the big optical telescopes. Radio telescopes pick out many more remote galaxies, but they are untypical ones. Just what happens at the limits of detectability – optical and radio – is a vexed and pregnant question concerning the origin and fate of the universe, which can be postponed till the next chapter.

We have universe enough for the moment. Look at the dots on the back of the envelope, and visualise each one as a multitude of stars, amid which astronomers on a few thousand planets may catalogue the Milky Way itself as a commonplace spiral galaxy. In 30 million years or so, our neighbour in the Virgo cluster may watch with interest that exploding star in our Galaxy, which *we* saw in 1054 AD.

There is one kind of measurement that, more than any other, dominates current astronomy in its study of distant objects, and also tells us the most important feature of the universe. With the exception of the nearest galaxies, all the distant objects show a persistent and growing effect in their light: it is redder than it would be if we were sitting in, or close to, the objects. More precisely, different elements in the stars of the galaxies absorb or emit light of well-defined wavelengths but, in the light as we see it, from far away, all these wavelengths are somewhat longer than they should be. The reason accepted by most astronomers is that the galaxies are all moving away from us at high speed so that the light waves are, as it were, stretched by this relative motion. The greater the speed of the galaxy, the greater the stretching, or reddening.

*Redshift – the key measurement of modern astronomy. Spread out into its coloured spectrum by a grating, the light of stars and galaxies shows dark lines (such as H and K lines due to absorption of light of particular wavelengths by calcium atoms). In a distant, receding galaxy these lines are shifted towards the red end of the spectrum. The redshift is sometimes expressed as the percentage increase in wavelengths.*

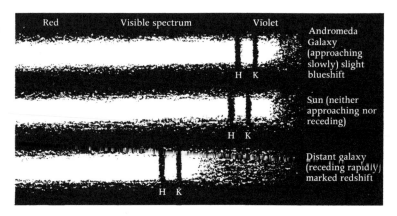

Red    Visible spectrum    Violet

Andromeda Galaxy (approaching slowly) slight blueshift

H  K

Sun (neither approaching nor receding)

H  K

Distant galaxy (receding rapidly) marked redshift

H  K

One exception to the pattern of movements is the nearest spiral galaxy in Andromeda, M31, which seems to be coasting gently towards us at a speed of 125 miles a second. Its light is blueshifted, because all the wavelengths are slightly compressed, but that represents only a random motion within the local group. There is, incidentally, little risk of a collision between M31 and our Galaxy, because the apparent approach is in fact due to the motion of the Sun and Earth in orbit around the centre of the Galaxy.

Otherwise, the universe as a whole seems to be expanding, or exploding. From measurements of the redshift, related by a series of steps to distances astronomers can measure by triangulation, we learn both that distances in the universe are immense and that they are growing in a systematic way. All the spaces between all the galaxies are increasing.

*A group of five galaxies with unusual connecting clouds between them (NGC 6027 in Serpens). Such groups represent another disorderly form of galaxies. (Mt Wilson & Palomar Observatories 200″)*

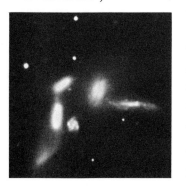

## Tell-tale variations

The realm of galaxies has great qualities of serenity, beauty and apparent timelessness, as some of the accompanying photographs may convey. It is a pity to spoil the picture, but astronomers today are fascinated and perplexed by the discovery that, among the countless galaxies, there are many in a state of extreme disorder. A galaxy as a whole can explode. One graphic case is M82 (see colour illustration between pages 40–1); another is M87, a galaxy that has plainly ejected huge jets of matter. Halton Arp of Palomar has an *Atlas of Peculiar*

*Galaxies*, which includes many fantastically distorted objects. Bracketed with the exploding galaxies are the quasars. These are objects comparable in mass and distance with the ordinary galaxies, but much smaller and yet greatly exceeding them in brightness. The problem at present is not so much to imagine why such large masses come to be involved in a disaster but to explain how, given their participation, they can survive and release energy long enough and fast enough to account for what we detect, considering how devastating the force of gravity must be in such concentrations of mass.

Several of the telescopes of the Lick Observatory, perched on Mount Hamilton 50 miles from San Francisco, serve in Tom Kinman's programme for measuring the brightness of distant objects in the sky, including the big 120-inch telescope. Kinman is an authority on variations in brightness that occur in some of the galaxies and quasars, and that special interest requires him and his colleagues to keep watch on these objects. Repeatedly they photograph them among the stars lying in their foreground; the astronomers use the big telescope to check the brightness of those foreground stars, so as to fix, by comparison, the brightness changes in the objects of interest.

The tale these variations tell is of great violence, greatly exceeding

*One of the galaxies most conspicuously in distress is M87, which has thrown out a huge jet of matter (NGC 4486, in Virgo). This short-exposure photograph picks out the jet and the bright centre, but not the fainter outer regions of the galaxy. (Lick Observatory 120″)*

*Tom Kinman, a British astronomer working at the Lick Observatory, who investigates variable galaxies and quasars.*

the waxing and waning of an exploding star and certainly involving matter equivalent to vast numbers of stars. The variations also say something very important about these objects: that they are compact. The argument is simple, though complicated in detail by gravitational effects of high-speed motion of the objects and their component material. Just as the high frequency of a pulsar's beat, related to the speed of light (see page 36), forces astronomers to think that the active regions of pulsars must be small, so the fluctuations week by week in some of the much more distant and more powerful objects speak of relatively small size, too. If a source can change markedly in a couple of weeks, it cannot be much more than a few light-weeks, or perhaps light-months, in diameter. Our own Galaxy is 100,000 light-years in diameter – yet these small, distant objects pump out far more energy than the Galaxy does.

No one can yet explain with confidence what goes on out there among the exploding galaxies and quasars. But investigations like Kinman's are likely to hasten the correct explanations, or at least rule out poor ones. Only a minority of objects in the distant realms of the galaxies show peculiarities of form and brightness, and only a minority of this minority vary in brightness from week to week. But, as Kinman says, 'The variations help us to see something like an experiment in progress – one which can perhaps show us how the energy is released'.

Kinman's favourite object is the quasar 3C345. It waxes and wanes in brightness over a period of about a year, but every twelve weeks or so it abruptly doubles in brightness in the course of a few days, and as quickly declines again. The light in the slow and especially the fast additions to the brightness of 3C345 shows evidence of being generated in the presence of strong magnetic fields. In connection

*A 'peculiar' galaxy (NGC 2623 in Cancer). It looks misshapen and is a strong source of radio noise. (Mt Wilson & Palomar Observatories 200'')*

*120-inch telescope at the Lick Observatory, ranking second only to the 200-inch telescope on Palomar Mountain.*

with the BBC-PBL programme, the Lick Observatory has prepared the first movie (much speeded up!) showing the fluctuations in a quasar, using photographs of 3C345. That is just as well, because otherwise quasars, being quasi-stellar in appearance, look very unimpressive.

Another quasar that Kinman watches is 3C279. It shows the same general pattern of change as 3C345, but its short-term variations are much more marked, with a sixteen-fold increase in light output – again associated with strong magnetism. J. B. Oke at Palomar has seen this object double its brightness in 24 hours. Equally flashy behaviour characterises 3C446, a distant object studied by several observatories in the USA, and the British observatory at Herstmonceux.

Everyone knows that quasars are funny things. The name 'crazy stellar objects', attributed to the young daughter of two distinguished astronomers, Margaret and Geoffrey Burbidge, is as apt a malapropism as one could wish for. These fluctuations in brightness, as Kinman says, give 'experimental' information on their nature. But, in another case, they also compound the mystery, because the object in question is not a quasar. It is a galaxy, albeit an unusual one. Known now also as a variable radio source, 3C120 was spotted by Harvard astronomers as long ago as 1940, but they thought it was a variable star. It is really a galaxy with an unusually bright centre and in 1968 Kinman has watched it lose half its brightness in the course of seven weeks.

### No traffic accident

With hindsight, we can say the present mystery of exploding galaxies began 18 years ago, in 1951. That summer Graham Smith, then at Cambridge, measured the position of the second brightest radio source in the sky, Cygnus A, precisely enough for the distinguished American astronomer, Walter Baade, to train the new 200-inch Palomar telescope on the spot, and take photographs. Baade saw there was something unusual as soon as he developed the negatives. Right in the middle of the picture was what looked like a pair of galaxies pushed together, like a dumb-bell.

By misleading chance, the astronomer who first looked at this picture had for years speculated about the chances of galaxies colliding with one another while they moved about the universe at high speed. Baade quickly convinced himself, and then his colleagues, that

*This galaxy seems to be tearing itself apart (NGC 4038-9 in Corvus). Compare it with the galaxy of Cygnus A, mistaken at first for a pair of galaxies in collision. (Mt Wilson & Palomar Observatories 48″)*

*Cygnus A. As the radio map shows, a galaxy far away in the universe has expelled two huge radio-emitting clouds (Mullard Radio Astronomy Observatory). At the centre, ringed in map, is a misshapen galaxy. (Mt Wilson & Palomar Observatories 200″)*

Cygnus A was a pair of galaxies in collision – a traffic accident on a stupendous scale. The very strong radio emission of this and other distant radio sources was then to be explained by the long drawn-out crash of the gas clouds in the two galaxies. For about ten years, the idea was widely fancied, although even in the 1950s some leading radio astronomers, including Hanbury Brown at Jodrell Bank and John Bolton in Australia, were increasingly mistrustful of this explanation of the radio galaxies.

Cygnus A is more than 500 million light-years away and yet it appears as a stronger radio source than the Crab Nebula, created by an exploding star a mere 6,000 light-years away. You need the equivalent of ten billion Crab Nebulae to account for Cygnus A. A very high-speed collision between two galaxies might conceivably supply the energy of the radio galaxies, yet repeated attempts to prove relative movement in the visible objects failed. Moreover, as radio telescopes came to be used in more precise ways, it turned out that the radio-emitting region of Cygnus A was twenty times bigger than the visible object. Again a dumb-bell pattern appeared, but the two parts of the radio source were spread on either side of the visible object and did not coincide with it at all. A similar pattern was found for other radio galaxies and when, in some cases, the radio source overlay the corresponding visible object, that object would look like a single galaxy in distress, rather than a pair in collision.

## Blue star blues

The years 1960-2 saw ideas about radio sources put back in the melting pot. The simple fact that some of the radio sources scintillated – twinkled like stars – was a sign that they must be rather small. Henry Palmer of Jodrell Bank used the 250-foot radio telescope and

3C 273. A quasar seen in the
200-inch telescope. This object is
very far away, but the jet is
reminiscent of that in M 87
(see photograph on page 85).
(Mt Wilson & Palomar
Observatories)

lesser telescopes 36 and 70 miles away as a means of measuring the
sizes of small radio sources (see page 101). He found that several of the
sources he looked at were very small indeed – less than a fiftieth of
the diameter of typical radio galaxies. In California, John Bolton and
J. Matthews pinpointed one of these sources, 3C48, and identified it
with a blue star. It was the first star-like radio source, contrasting
with the smudged, cloudy objects of all previously known sources.
Three other radio sources were identified with similar blue stars. It
was naturally supposed that they lay in our own galaxy. Optical
astronomers examined their light, but could make no sense of the
pattern of strong wavelengths.

In November 1962, the Moon happened to pass in front of a fifth
small radio source, 3C273, and, in Australia, Cyril Hazard and his
colleagues took the opportunity to use the new Parkes radio telescope
to fix very precisely the time at which the edge of the Moon blocked
the radio signal. Thus they were able to show that 3C273 consisted of
two very small patches and to pass precise positions to Maarten
Schmidt, a young Dutch astronomer at Palomar who was interested
in the star-like radio sources. He found once again a blue star, but this

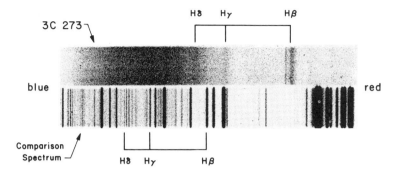

*Redshift of 3C 273. It was the marked displacement in wavelength in a recognisable pattern of lines (labelled Hδ, Hγ, Hβ) that led to the realisation that quasars lay at immense distances.*

time with a faint jet protruding from it; the twin radio sources coincided with the star and the top of the jet.

Schmidt examined the composition of light from 3C273. Again he found a pattern of strong wavelengths, which he could not understand. His colleague, J. B. Oke, detected another strong emission in the invisible infra-red. But still the pattern made sense to no one. Then, several weeks later, on 5 February 1963, Schmidt experienced one of those flashes of insight that are the supreme reward of the scientific life. Puzzling once more over the spectrum of the light of 3C273, because he had to write a report on it, he noticed that the basic pattern was just like that of the light of hydrogen gas, but with all the wavelengths greatly stretched. In other words, each component was shifted towards the red end of the spectrum.

The idea of such a redshift was not strange to Schmidt. On the contrary, he knew it well as the means by which distances of the farthest galaxies were measured. But it was a remarkable thing to encounter in a star; stranger still that the extent of the redshift put the 'star' far out among the galaxies – 1,500 million light-years away. It seemed to Schmidt an almost incredible and also a terrible discovery, because it meant that 3C273, looking no bigger than a star, was shining as brightly as 200 galaxies.

Schmidt yelled down the corridor to his colleague Jesse Greenstein, who quickly pulled out his own records of the light from the first of the radio-emitting blue stars, 3C48, and examined them again. If those lines were redshifted . . .? Yes, a previously baffling pattern fell into place, given an even greater redshift than for 3C273, corresponding to a distance of 3,600 million light-years.

Thus did the quasi-stellar radio sources, or quasars, come to be recognised for the fantastic objects that they are. Six years later the excitement of Schmidt's discovery has scarcely abated, while the dismay they arouse grows as time passes and a satisfactory explanation

*Jesse Greenstein, who worked with Schmidt on the day of his discovery, now gives his attention to exploding stars in distant galaxies.*

for them is still lacking. In 1969, quasars are being spotted in growing numbers in radio observations. From a discovery at Palomar in 1965, Allan Sandage believes that millions of blue stars, previously supposed to be faint objects in the halo of our Galaxy, are also quasars – but 'quiet' ones that are not registered by the radio telescopes. Conversely, the radio astronomers pick up small radio sources from positions where no plausible correspondence can be found with any visible object; these 'empty-field' sources are almost certainly quasars too far away to be visible at all, but still detectable by radio.

The problem of the enormous radio power, already encountered in the radio galaxies, is compounded in the quasars by the need to generate intense light, too, and to squeeze all that energy out of a very small object. The riddle would be easier to answer if the quasars quickly burned out, as do exploding stars, yet known quasars can be seen burning brightly on old sky photographs taken 80 years ago. Moreover, in 3C273, the length of the jet indicates that it was expelled at least 150,000 years ago!

When radically conflicting theories abound, it is a sure sign of ignorance, and this state of affairs has persisted for several years among students of the quasars. The weight of opinion at present falls firmly on the side of Schmidt and Greenstein's initial opinion that the quasars are at very great distances. Here is just a selection of explanations offered for the great energy of the quasar – all of them proposed by reputable astronomers and physicists:

1. A great mass of gas collects under gravity to form a new galaxy. At the outset it makes large numbers of massive stars, which become supernovae and explode in quick succession, at a rate of several a day.

2. A great mass of stars collapses together, and the stars collide, creating vast numbers of supernova-like explosions.

3. A great mass of gas collapses under gravity and goes on collapsing indefinitely. This mechanism could release a great deal of energy, but in a short while the energy would be unable to escape – if present laws of gravity are correct.

4. Clouds of 'anti-matter' exist in the universe and react with matter in quasars. Anti-matter is known from laboratory experiments but there is no other evidence for its existence in important amounts in the universe. Anti-matter is very like ordinary matter; in eerie and violent fashion, however, matter and anti-matter mutually annihilate themselves on contact, releasing intense energy.

5. There is a new but so far unspecified source of energy in the universe, of which the quasars are the most dramatic manifestation.

6. The quasars are remnants of the original stuff from which the universe was formed at the time of the Big Bang.

7. The quasars are newly created matter – cosmic taps whereby the stuff of the universe is replenished.

That the Harvard astronomers in 1940 mistook the galaxy 3C120 for a variable star is not only a forgivable error but a highly suggestive one. It is one of a class of galaxies known as seyfert galaxies, after their discoverer, Carl Seyfert. They have centres so bright in comparison with their surroundings that, in short-exposure photographs only the centres are visible – and they look like stars. About one galaxy in a hundred is a seyfert. As these bright-eyed galaxies become better known, a growing number of astronomers are convinced of a link between quasars and galaxies. Quasars may be galaxies with exceptionally bright centres, so that the regions outside the centre are invisibly weak in comparison. Some quasars (3C48 in particular), look slightly fuzzy, as if they are galaxies. But nearly all the known galaxies, even the brightest seyferts, have less than a thousandth of the brightness of the quasars. That leaves a big gap in brightness to be filled if there are family connections between quasars and galaxies.

*I Zwicky 1, the brightest galaxy in the sky which, for Wallace Sargent, is the 'missing link' between quasars and galaxies. This is a negative picture, of the kind with which astronomers work, to see fine detail. (Mt Wilson & Palomar Observatories)*

Odd-looking galaxies, especially small, compact and bright ones, hold the special attention of Wallace Sargent, a British astronomer working at Palomar. He finds that a strange object, known as I Zwicky 1 after the astronomer who first catalogued it, is the brightest galaxy known. 'Its centre is 100 billion times brighter than the Sun,' Sargent says. 'I have suggested that this object is really the missing link between galaxies and quasars.'

Sargent is also interested in another strange formation, designated Vorontsov-Velyaminov 172 after the Soviet astronomer who, in the 1950s, looked for odd galaxies in Palomar sky photographs, and found them in abundance. VV172 is actually a row of five compact galaxies, that look plainly as if they are in company. Four of them are, but Sargent, comparing their speeds, finds that one is the odd man out – moving away from the others at a speed of at least 15,000 miles a second. 'What can have happened here,' Sargent asks, 'except some tremendous explosion?'

The odds against the fast-moving member of VV172 being a more distant galaxy, that just happens to lie in the same direction as the other four, are about 5000 to 1. Sargent's discovery lends support to the idea of Victor Ambartsumian that new galaxies can be created by explosions in the centre of old galaxies. In the case of VV 172, the explosions may have occured just a few million years ago. This group, though an extreme case of a galaxy flying apart from its companions, is not unique. Earlier, Zwicky found that one of a group of three galaxies, linked together by luminous streamers, was moving away from the others at 5000 miles a second. Then Margaret and Geoffrey Burbidge checked another group – known as Stephan's quintet – and

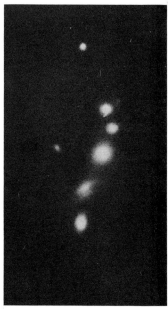

*VV 172 and the odd man out. The second galaxy from the top is travelling away from its companions at immense speed. (Mt Wilson & Palomar Observatories)*

*Exploding Galaxies*

*Stephan's Quintet. Here, too, one of the galaxies is breaking away from its partners at high speed. The runaway is the spiral galaxy at lower left. (Lick Observatory 120")*

it turned out that one of these five galaxies was parting company from the others at similar speed.

Some theorists see, in such vast explosions, projecting massive bodies as shrapnel in different directions, a possible explanation for the high speed of recession of the quasars, which would bring them much nearer to us than does the simple interpretation of their red-shift in terms of the general expansion of the universe. They could be blown out from explosions in relatively nearby galaxies. One snag is that nobody has found a blue-shifted quasar — one travelling towards us — and it would be too much of a coincidence if they all happened to be pushed away from us in the supposed explosions. The only plausible version of this idea is that they were expelled from

*A seyfert galaxy. NGC 1275 is one of the most remarkable galaxies observed by astronomers. Besides showing marked variations in brightness it is also a powerful source of infra-red rays. (Mt Wilson & Palomar Observatories 200″)*

an explosion in our own Galaxy or one very close to us. In any case, the problems of explaining the vast concentrations of energy in the quasars are not eased by a supposition of this kind. The total energy, to be sure, can be less, but the possible diameter of the object is proportionally smaller, too.

Sargent's view that the seyfert galaxies and the quasars are similar objects showing different degrees of violence remains very persuasive. And, as we shall see later, the seyferts turn out to be even more energetic than they appear to be, in optical and radio observations, when examined at another wavelength.

## Galaxies in distress

Radio astronomers, too, see resemblances between galaxies and quasars. Sir Martin Ryle tells the most coherent story here, based on years of careful mapping of radio galaxies and quasars by his group at the Mullard Radio Astronomy Observatory, Cambridge University. The dumb-bell form of radio galaxies such as Cygnus A, the first to be discovered, is the key to his interpretation. Most radio galaxies have such a double form, with a pair of large clouds emitting strong radio waves, often with an odd-looking galaxy visible between them. This appearance suggests that radio-emitting plasma clouds, containing high-energy electrons and magnetic fields, have been expelled in opposite directions from a central explosion in the galaxy. The wide separations of the clouds, combined with the astronomers' belief that these clouds can go on broadcasting only for a few million years before they run out of high-energy electrons, suggests that the clouds travel at practically the speed of light.

Some quasars, too, have a double structure in the radio maps, though on a smaller scale. This possibility was known from the early days of quasars, with the discovery of the double radio source in the case of 3C273. On the other hand, another of the earlier quasars to be discovered, 3C48, appears to have a single, small source of radio waves.

Ryle says it is very difficult to draw a sharp dividing line between quasars and radio galaxies. Some objects that look like quasars, by radio, prove to be visible galaxies; others that look like radio galaxies, by radio, turn out to be quasi-stellar in the optical telescopes. It also stands to reason that what we now see as widely-spread clouds in a radio galaxy must have been much closer together at the early stages of the explosion. Conversely, the immense release of energy in a quasar suggests that it may eventually expand, to make something like a radio galaxy. Moreover, although the quasars, by their extreme compactness, are the more astonishing objects, the release of energy represented in a radio galaxy is quite comparable with that of a quasar.

*This radio source, 3C47, has the double cloud of a typical radio galaxy, but the object in the centre (+) appears to be a quasar. Associations of this kind suggest to the Cambridge radio astronomers the evolutionary link shown right. (Mullard Radio Astronomy Observatory)*

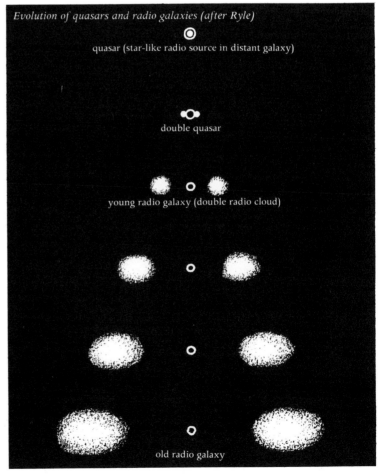

*Evolution of quasars and radio galaxies (after Ryle)*

quasar (star-like radio source in distant galaxy)

double quasar

young radio galaxy (double radio cloud)

old radio galaxy

All these arguments Ryle deploys to say that quasars and radio galaxies are simply the same kind of explosive object, at different ages. He and M. S. Longair put a time scale to the changes as follows. There is an explosion in the centre of a galaxy. The release of energy is equivalent at least to $10^{61}$ ergs (10 followed by 60 noughts or roughly 100 billion billion billion billion H-bombs). For the first 30,000 years the object is a quasar and the radio emission is concentrated at fairly short wavelengths, because the longer radio waves are absorbed within the source itself. Thereafter, and until about 300,000 years after the explosion, there may be ambiguity about whether the object is a quasar or a radio galaxy. But at E-day plus 100,000 years or so, the plasma clouds responsible for the radio emission expand out of the galaxy into the near-vacuum of intergalactic space. The intensity of the radio emission decreases rapidly and becomes weak after 3 million years, though the clouds continue to fly apart and swell. Galaxies may explode more than once. Some radio galaxies show several 'puffs' at various distances from the centre. These can be interpreted as signs of a series of explosions within the galaxy.

One of the various explanations ventured for the quasar explosions (see page 92) finds support, according to Ryle, in observations of what he calls a quasar in slow motion. This is another bright-eyed seyfert galaxy, NGC1275. As a radio emitter (3C84), it is much weaker than a quasar but there is a very compact source in the centre, which could be a large star exploding; the radio emission from its immediate surroundings, Ryle estimates, could be accounted for, if such a supernova had occurred every few years. Margaret Burbidge, the optical astronomer, has seen gas streams emerging from NGC1275; radio observations also pick up emissions from further afield. The latter can mean only that the galaxy has been bubbling away as at present for more than five million years, releasing altogether as much

*Centaurus A. At least two explosions have occurred in this strange-looking spherical galaxy (NGC 5128 in Centaurus, Mt Wilson & Palomar Observatories 200''). The radio map shows one widespread pair of radio-emitting clouds, but there is also another, much smaller pair close to the galaxy, corresponding to a more recent explosion.*

energy as a quasar, but much more slowly. All of this, supported by optical evidence from Russian astronomers of a succession of supernovae in NGC1275, lends credence, in Ryle's view, to Geoffrey Burbidge's early idea that quasars proper may be produced by a chain reaction of supernovae, in which one exploding star triggers explosions in others, so that a million supernovae occur in the same galaxy, within a hundred years.

The rate at which the energy is released, as well as what it amounts to, can have a big effect on the appearance of a galaxy in distress. A rapid release of energy makes the object bright in both optical and radio observations. A somewhat slower rate could leave the radio emission strong but make the galaxy unimpressive or undetectable as a visible object. Ryle thinks that the quiet quasars discovered by Allan Sandage, very bright but unrecorded by radio, may be galaxies in which relatively mild explosions occur, with long intervals between them.

### Seeing red

There is a consensus among both optical and radio astronomers who see the quasars, the radio galaxies and the galaxies visibly in disarray as being related objects – various conditions of exploding galaxies. The observations and arguments marshalled by Kinman, Sargent, Ryle and others are impressive. But it would be wrong, in this dramatic period of astronomy to suggest that the story is told, once and for all. For those who beg to count themselves out of the consensus, there is still room for manoeuvre.

According to some, the variations in brightness of quasars could be caused, at least in part, by opaque clouds passing between the quasars and ourselves. More significantly, part of the redshift in quasar light could be due, not to great speed of recession, but to the effects of strong gravity – with the light losing energy in climbing away from an extremely massive body, and so shifting towards the red. There are even suggestions that the redshifts are deceptive because the laws of atomic physics, governing the generation of light, are different in the quasars. Leading optical astronomers find all such ideas horripilant, as they cast doubt on the classical interpretation of the redshift, as a measure of both velocity and distance in an expanding universe. 'If that went,' Jesse Greenstein says, 'it would be worse than the fall of the Roman Empire'.

Quasars are enigmatic objects that give little away about their real nature – just a few lines in a spectrum, a blob or two on the radio map, a variation in brightness. The scant information available is contradictory or confusing. Some quasars show several *different* degrees of redshift. And Geoffrey Burbidge in San Diego frets about a strange coincidence in the redshifts of the quasars and other explosive objects. The natural expectation would be of a random assortment of redshifts among the quasars, from low to high values, as there is for ordinary galaxies. But it turns out that they cluster around preferred values, notably around that corresponding to an apparent speed of recession of 150,000 miles a second, and a supposed distance of 8 billion light-years.

His suspicions being aroused by this coincidence, Burbidge looks closer and finds clustering in many of the quasars around redshifts that are multiples of 11,000 miles a second in apparent speed of recession. Does that, he asks, mean that the quasars are dotted in a regular pattern through the Universe? That would be preposterous, or so Burbidge answers himself; instead he looks to some other physical effect, such as gravity, to produce the redshifts. Another possible interpretation of Burbidge's patterns of redshifts is that quasars are produced at rhythmic intervals — there are certain preferred times for their formation.

One contrast between ordinary big galaxies on the one hand, and seyfert galaxies and quasars on the other, gives satisfaction to dissenters from the view that the big redshift of the quasars is due to enormous distances. Halton Arp shows that the redshift increases more quickly, as the explosive objects grow fainter, than it does for the ordinary galaxies. This very provocative result may tell of a real natural effect, or it may be that the explosive objects selected for comparison are unrepresentative.

### A bee at 4000 miles

An obvious way to try to settle arguments about the quasars, and to channel speculation about how they work, is to secure clearer pictures of them. In the optical telescopes they look like pinpoints — with the notable exception of 3C273, which is a pinpoint with a bit of the pin sticking out from it—the now famous jet. The radio astronomers have a better chance of seeing some size and shape in the radio-emitting regions of the quasars. To that end they are pushing their techniques to the limit, and in particular they adopt an idea first propounded for optical astronomy just 100 years ago. To explain it and show its significance, I must digress into some simple principles of telescopes, and a little bit of history.

The power of a telescope to resolve, or distinguish between, two nearby points in the sky depends, at least in theory, on the diameter of the telescope measured in *wavelengths* of the radiation it is detecting. The bigger the diameter, or aperture, the greater the resolving power of the instrument, and the smaller the objects it can distinguish. The pupils of our eyes have a typical diameter of about 5,000 light wavelengths, and we can, for example, see a bee at distances of up to about 50 yards – corresponding to an angle from the eye of one arc-minute. The aperture of the 200-inch Palomar telescope, for comparison, is about 10 million wavelengths, and in theory it can separate objects three hundredths of an arc-second apart. In practice, the distortions of the atmosphere usually limit the resolving power to about one arc-second, or only about 60 times better than the human eye; it corresponds to spotting the bee at rather less than two miles. The large aperture at Palomar therefore serves more to gather light from faint objects than to show bright objects with exceptional clarity.

The finest measurements of visible objects are made with a special-purpose instrument, operated at Narrabri, New South Wales, by R. Hanbury Brown. He is a former radio astronomer from Jodrell Bank who has turned to optical astronomy, in order to check the sizes of some individual stars. Their distances are known, so the important measurement is the very small angle of the sky that each one fills. Then a simple sum will give its true diameter.

Hanbury Brown's set-up, in the hot outback of New South Wales, is reminiscent of a radio observatory. There are two big mirrors which can be spaced on a circular railway, at different intervals up to 600 feet apart. This combination gives an 'aperture' of up to 400 million wavelengths, with a resolving power of better than a thousandth of a second of arc. That is like seeing our friend the bee at a distance of 4000 miles.

The dodge that Hanbury Brown uses was proposed in France just 100 years ago. Hippolyte Fizeau was a wealthy Parisian whose hobby was the study of light. He made the first direct measurement of the speed of light, using a chopped beam travelling from Montmartre to Suresnes and back. Fizeau was also the first to show how to find the speed of a star moving away from us or towards us, by the redshift or blueshift of particular wavelengths in its light; applied to galaxies and quasars, it is one of the most important techniques in astronomy in the 20th Century. Anyway, in 1868, Fizeau saw a new consequence of the fact that light is made up of waves: it should enable astronomers to measure the incredibly small angle made by the face of a star, seen from the Earth.

A modern analogy illustrates the principle. Imagine two television cameras looking at the same scene in a studio and both views being fed simultaneously as a 'mix' on to the viewing screen. Try as they may, the technicians will not be able to overlay the two pictures perfectly. The result may not be too bad, if the cameras are very close together, but separate them and the confusion will be terrible. There is one exception: if the studio is dark except for a single small light-bulb, the two views of that light-bulb can be superimposed almost perfectly, however far apart the cameras are.

In the astronomical case, a double telescope replaces the cameras, a star replaces the scene. The great precision comes by seeing, not whether the pictures of the star can be matched, but whether its light waves in the double telescope can be superimposed. The bigger the star, the closer the spacing of the double telescope must be, to avoid confusion of the light waves. So the size of the star can be measured by progressively widening the spacing, and seeing where confusion begins. The useful paradox of Fizeau's proposal is that the smaller an object is, the easier it is to see it free from confusion.

A nice idea, but very tricky in practice. It was taken up by the great American physicist, Albert Michelson, when he was 38 years old; not until he was 68 did he succeed in making it work. That was at Mount Wilson in 1920, and no one has managed to improve on the measurements of a handful of nearby stars, made at the time by F. G. Pease, until Hanbury Brown's current work. He modified the techniques in a way too subtle to explain in detail, but the essential idea remains, of moving two telescopes apart until confusion sets in.

Even the 1000-foot Arecibo radio telescope, working at, say, 70 centimetres wavelength, has an aperture of only 430 wavelengths. Its resolving power is ten times worse than the human eye's – it cannot distinguish objects less than a sixth of a degree apart. As a result, the view of the objects in the sky is blurred. By ingenious combinations of telescopes, like the three-dish, one-mile telescope at Cambridge (see page 123), radio astronomers can greatly improve their vision, but only narrowly to beat the unaided human eye in resolving power. At this scale of operations, it therefore seems as if the optical astronomers must always have the sharper view of the universe – unless the radio astronomers build telescopes hundreds or thousands of miles in diameter. But that is, in effect, just what the radio astronomers are doing.

The first step was taken at Jodrell Bank around 1960 – and Sir Bernard Lovell tells me he regards this work as the most important single contribution of his observatory. At some tens of miles distant from Jodrell Bank, a smaller dish would be set up and it would relay received signals to Jodrell by Radio link. With such double telescopes Henry Palmer and his colleagues looked for the onset of confusion in the signals from various radio sources, as the separation increased. Many sources duly revealed their sizes, but some stubbornly did not, even when the separation was increased again and again. This was the first hint of the existence of very small radio sources – the quasars (see pp. 89ff).

Nowadays the Jodrell Bank radio astronomers are working routinely with baselines of up to 85 miles, in conjunction with a radio telescope of the Royal Radar Establishment at Defford. By this means they continue the investigations of fairly small radio sources. To measure the smallest, much greater baselines are needed, but the radio link becomes increasingly difficult to stretch.

This limitation has now been overcome, by radio astronomers on the other side of the Atlantic, to create telescope combinations that equal, in their resolving power, Hanbury Brown's optical instrument, and reveal finer detail in the quasars than the great Palomar telescope can ever be expected to show.

*Telescope as big as the Earth*

It is a long drive from Ottawa via Pembroke, Ontario, and the site of the Algonquin Radio Observatory corresponds very exactly with the popular image of old, wild Canada. It stands amid the forests of a national park, by Lake Traverse, and there is even a lumber camp nearby to complete the picture. But there is nothing old-fashioned about the science. The pride of the Canadian radio astronomers is a 150-foot dish, very like a scaled down version of the Australians' 210-foot dish at Parkes, and figured for accurate working down to three centimetres radio wavelength. When it came into operation, the Canadians promptly started a revolution in radio astronomy under the banner of 'very long baseline interferometry', early in 1967. Across the border, American radio astronomers began getting results, using the same VLB principle, at almost the same time. Two years later, this VLB technique is giving important new information about the quasars. It combines observations from radio telescopes at great distances apart, to reveal details not otherwise distinguishable.

Norman Broten and his fellow radio astronomers in Canada extend the possible baselines to the limit set only by the size of the Earth. Their big step was to replace the radio link used by Henry Palmer at Jodrell Bank with high-capacity tape recorders (surplus from the Canadian Broadcasting Corporation) at the two ends of the baseline. The signals they record would be meaningless if they did not add, to the recordings, regular time pips from an extremely accurate atomic clock at each observatory. The most accurate form of such clocks would not lose or gain a second in a million years. Before each series of observations, the two clocks are brought together and adjusted so that they run at exactly the same rate. When the tapes are full of radio records, these, too, are brought together at Algonquin and blended: the time pips ensure there is continuous matching from moment to moment. Provided the telescopes were looking at a small enough radio source there is no confusion. A pattern appears, just as if the two telescopes had been linked at the time of observation. The

system was first tested between Algonquin and Ottawa, and then used in earnest between Algonquin and Penticton, in British Columbia on the other side of the continent.

American radio astronomers from several observatories have carried out extensive work with their VLB system. It is very similar although, as they admit ruefully, the Canadians' is much more economical of magnetic tape and computing time. The Americans have operated on several baselines from Green Bank (West Virginia) or Haystack (Massachusetts), to Hat Creek (California), Arecibo (Puerto Rico) and Onsala (Sweden).

The first results of the very long baseline observations of quasars are now becoming available, and structure emerges at last in these cryptic objects. The American group reports that half the radio energy from the star-like object in 3C273 comes from a region about 40 light-years in diameter, but there are other regions much smaller in size. (These figures assume that 3C273 is at the distance indicated by its redshift.) The Canadian group is amassing the information on this quasar and suspects it is cigar-shaped rather than round. Two other quasars, 3C286 and 3C309.1, are also showing some structure.

Closely related with the quasar story are the seyfert galaxies, notably the variable one, NGC 1275 (see pp. 97–8). VLB measurements show that the variable radio source in that galaxy is only about one light-year in diameter. Exploitation of very long baselines has only just begun and they will certainly have an impact on many branches of astronomy, far into the future. For example, new light will be shed on the birth of stars.

Nor is it only in astronomy that opportunities lie. The extraordinary accuracy of these radio telescope combinations is shown by some of

*Algonquin and Penticton. The first transcontinental radio-telescope combination was accomplished by Canadian radio astronomers using these two dishes.*

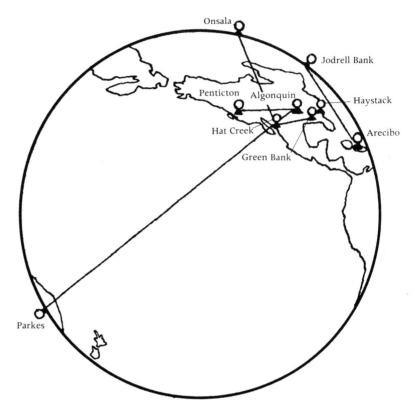

*Very long baselines. The principle of VLB operations and the main links used so far. The Algonquin-Parkes link gives a telescope virtually as big as Earth.*

the other uses they will have. Already, they are helping to test the fine points of Einstein's theory of gravity (see page 146). The VLB technique can also assist in mapping the Earth, because studies of the cosmic radio sources incidentally give the distance between the two ends of the baseline to an accuracy of a few inches. Such precision in Earth measurements opens up the possibility of checking directly the current theories of how the continents drift around like rafts on the face of the globe. The estimated rate of movement is only about half an inch a year – yet even this will be measurable by checking the intercontinental radio-telescope baselines over the course of a few years.

Meanwhile, the link-ups multiply, the baselines lengthen: Algonquin and Jodrell, Arecibo and Jodrell, Algonquin and Parkes. This last, in the Spring of 1969, just two years after the first Algonquin –Ottawa experiment, creates a baseline stretching across North America and the Pacific to Australia, a straight-line distance of 6,700 miles. It probably represents about the greatest physical separation astronomers will enjoy for a long time. By comparison the diameter of the Earth is 7,900 miles. Jodrell and Parkes would be an approximation to a pairing of telescopes at directly opposite points of the Earth's surface, but it would not work. The technique requires the two telescopes to be looking at the same part of the sky, yet only a very narrow strip of sky is simultaneously in sight from Jodrell and

Parkes, and then only with both telescopes trained on the horizon – which is not good for observing. Not until radio telescopes of high precision can be sent far out into space or established on the Moon will the mileage of the baselines be increased. In the Algonquin and Parkes hook-up, astronomers already have a telescope as big as the Earth.

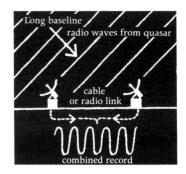

That does not rule out greater precision in observing radio sources with telescopes on Earth. What really matters is not the mileage between the telescopes but the number of wavelengths between them,

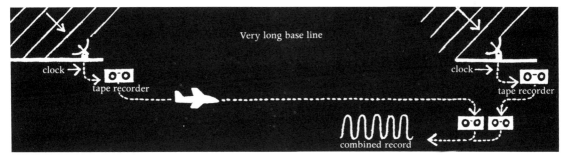

at the operating wavelength. Reducing the wavelength to a quarter effectively quadruples the length of the baseline, though that requires more accurately figured telescopes. Moreover, using varying baselines gives more detail and precision. We can look forward to magnificent enterprises of international co-operation, in which all countries owning great radio telescopes forge them, with tapes and computers, into one all-powerful instrument for looking at quasars and other small objects in the radio sky.

*Further freakiness – warm galaxies*

Frank Low's chief motive in flying with a telescope in a converted jet aircraft ten miles above California is his wish to confirm and extend his infra-red observations of galaxies. He and his colleagues have to be qualified in medical tests to fly at such heights, and they have to wear oxygen masks while they make their observations. The airfield is 800 miles from Low's home in Tucson, where he works in the University of Arizona. It would seem like a lot of trouble and inconvenience, but for Low's satisfaction that he is proving a technique of high-altitude astronomy better and cheaper for many purposes than balloon or rocket flights; and his belief that he is pioneering an exceptionally important new branch of astronomy. Infra-red astronomy began on the ground, but so great is the toll taken by the molecules of the atmosphere of most incoming infra-red rays from the

*Exploding Galaxies*

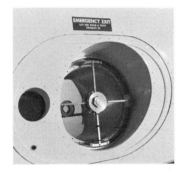

*Frank Low's telescope, fitted into the side of an aircraft (above), is manned by the astronomer equipped for flying to very high altitudes. See also colour illustration between pp. 136-7.*

universe that Low flies above most of the atmosphere for a better view. Low is by origin a physicist, expert in low-temperature instruments; he digresses into astronomy because he has developed sensitive detectors, cooled to very low temperatures and well-suited to picking up faint infra-red radiation from the sky.

Infra-red radiation was discovered, and infra-red astronomy began, in England in 1840 when Sir William Herschel put a thermometer beyond the red edge of a spectrum of the Sun, and found it became warm. For more than a century efforts to scan the sky in the infra-red were sporadic and inconclusive. Not until the 1960s did a number of American groups (ITT Corporation, University of Arizona, Caltech and Mount Wilson) begin work in earnest. Some of the discoveries, including the proto-stars and the centre of the Galaxy, have been mentioned earlier.

With a ground-based telescope at the Catalina Observatory near Tucson, Harold Johnson picked up more infra-red emission from the quasar 3C273 than anyone expected. Douglas Kleinmann and Low, still working with ground-based telescopes, found a remarkable amount of infra-red power coming from five of the bright-eyed seyfert galaxies – among them two old friends of ours, 'the slow-motion quasar' NGC1275 and the variable galaxy 3C120. The longer the infra-red wavelength, the stronger these objects appear,

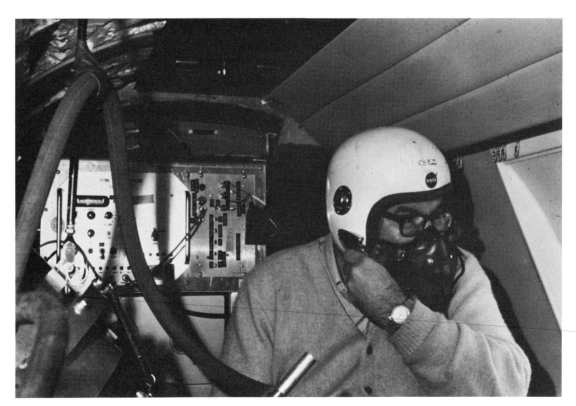

as if they should be expected to peak at a wavelength of about five hundredths of a millimetre. Low finds a similar trend in the emission from 3C273. But radiation at this expected peak cannot be well observed from the ground. High in the air, above most of the Earth's water vapour, it becomes possible.

Meanwhile, with Wallace Tucker, a colleague at Rice University, Low calculates the cosmic significance of the discovery of infra-red galaxies. If their conclusions are borne out by the ongoing observations, they transform the overall picture of energy in the universe. The seyfert galaxies, registering in the infra-red as extraordinary power-houses, pump out about a hundred times as much energy in the infra-red as they do in the form of visible light. Even if only one galaxy in a hundred is a seyfert, it follows that this small minority releases as much energy into the universe as all the other galaxies put together, but they do so in the infra-red. In other words, the invisible heat rays of the infra-red are as important in the energy budget of the universe as the light of the stars and galaxies.

Here, then, are galaxies that can be fairly said to generate more heat than light. The seyferts are bright, especially in their centres, but their predominant characteristic is warmth. And since seyfert galaxies are rising to such prominence in astronomical observation and debate, it is prudent to note that only twelve of them have been identified so far.

## Can it happen here?

After seeing all those catastrophic events in the centres of other galaxies, we may be forgiven for glancing somewhat nervously at the constellation of Sagittarius. Beyond it lies the centre of our own Galaxy. Could we, or our descendants, experience a mighty explosion in the heart of the Milky Way, turning our orderly spirals of stars into a spectacular wreck? If so, men might not live to tell the tale.

Life must be almost impossible in an exploding galaxy. The hazards to life on Earth of even a single nearby supernova were noted earlier (pp. 64–5). The blast of atomic radiation from a major explosion in the centre of a galaxy would be incomparably greater – so great in fact that, as a report by the Science Research Council in London puts it, 'it would strip the atmosphere of every planet like our own throughout the entire galaxy'. Nevertheless, to think what life would be like, assuming it remained possible at all, helps us to visualise the conditions prevailing in these exploding galaxies.

If there were a quasar in the centre of our Galaxy, it would not be very spectacular to see. Most of the light would at first be screened from us by the dust lying in the space of 30,000 light-years between

*Exploding Galaxies*

*Composite painting of Milky Way. The centre of our galaxy is roughly at the centre of the picture. (Lund Observatory)*

the quasar and us. Even if the dust were absent, the simple effect of distance would reduce the quasar from an object intrinsically 10,000 billion times brighter than the Sun to a point of light 4,000 times fainter than the Sun. It would, though, be a hundred times brighter than the Moon – lighting up our summer nights. Allowing for the dust, the quasar would be plainly visible in telescopes but not to the naked eye, until the radiation had swept a path for itself. Invisible

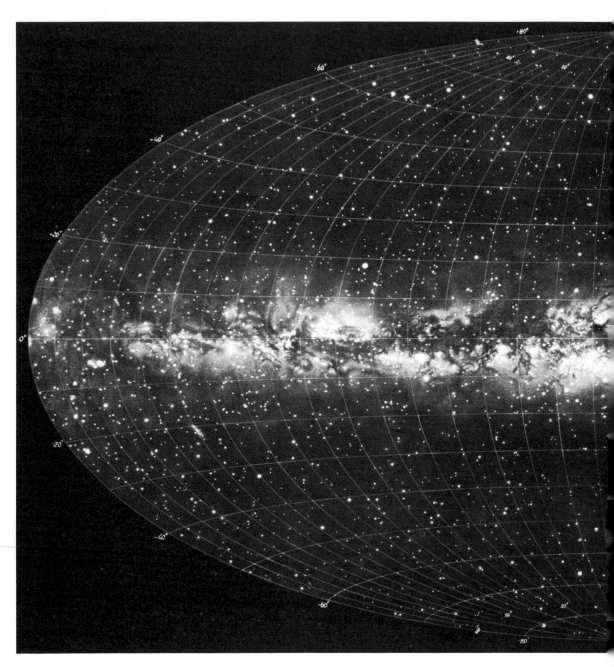

heat rays, radio waves, and atomic radiation, including the ghostly neutrinos, would be the main harbingers of the galactic disaster.

You can hardly be an astronomer in an exploding galaxy because, if by some miracle air bubbles out of the ground to replace that stripped by the radiation from the explosion, the atmosphere at night is lit by dazzling aurorae as atomic particles crash into it. The radio noise makes radio communications, never mind radio astronomy,

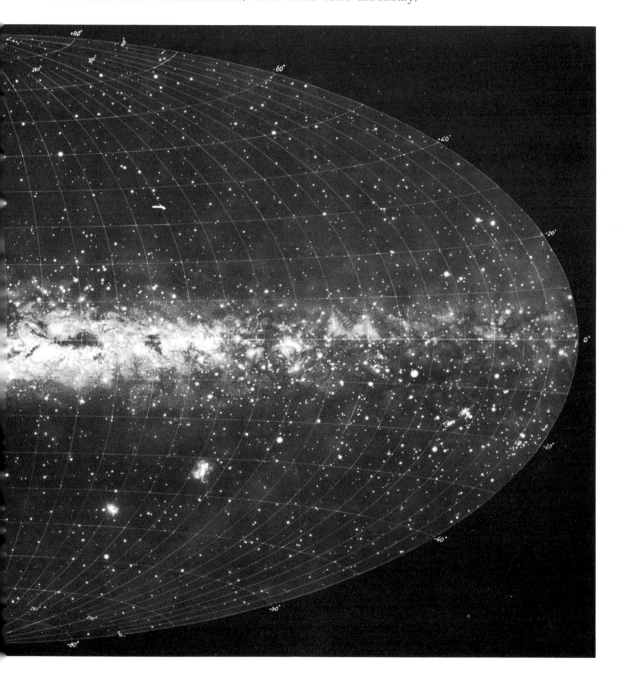

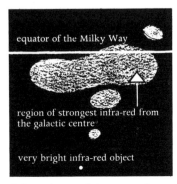

equator of the Milky Way

region of strongest infra-red from
the galactic centre

very bright infra-red object

*Infra-red map of the centre of our
Galaxy, showing strong emissions
both from the centre itself and
from a strange object nearby.
(California Institute of Technology)*

quite impossible. The seyfert galaxies, spluttering away and radiating strong heat waves, may be marginal cases for possible life. Although they are powerful sources of energy in the infra-red, if we lived in a seyfert the heat rays from the centre would add scarcely more warmth to the night than the Moon does.

A straightforward interpretation of the counts and distances of quasars is that the risks of galactic explosions were much greater a few billion years ago than they are now. In any case, major disasters in galaxies are probably the result of pathological defects, perhaps congenital – some galaxies may be simply too massive or condensed for safety. Our own galaxy looks fairly healthy in most respects.

Nevertheless, strange things do go on, in the middle of the Milky Way. Because of the dust, optical telescopes can see nothing. But in radio maps the centre of the Galaxy stands out clearly. It lies a few degrees south of the position you find marked on old star maps, which was the best guess of the optical astronomers. In gamma-ray maps, the centre of the Galaxy is by far the most prominent feature in the sky. In the infra-red, while not exactly conspicuous, the centre can be mapped in detail. In fact, the infra-red measuréments of Gerry Neugebauer and his colleagues at Caltech give one of the sharpest pictures so far, of our galactic centre. Radio mapping with the Haystack telescope has also given precise information.

Neugebauer sees a close similarity between the centre of our own Galaxy and the rather bright spot of the middle of the nearby galaxy in Andromeda. He ascribes the peak in the heat rays to a concentration of stars, ten million times greater than in the neighbourhood of the Sun. If we were in the centre of the Galaxy, the night lights would be far brighter than they are here, out in the suburbs.

Moreover, just a few light-years from the galactic centre, the infra-red detectors pick up a small, strong object. If we lived in that region, this object would be like a second Sun in the sky, turning night into day. Just what is the bright spot, 300,000 times more powerful than the Sun? It could be a mini-quasar, a supernova, a great condensation of matter, a single very bright star, or stars in collision. It, rather than the bigger, typically bright core of the Galaxy, is a symptom of possible incendiarism in the powder magazine of our galactic vessel.

Nor is that all. The Dutch astronomers who map the distribution of hydrogen in the Galaxy find a shell of gas moving outwards towards us at 33 miles a second, from the middle of the Milky Way. The simplest interpretation is that there was a big explosion in the centre of the Galaxy, about 30 million years ago. So perhaps it *can* happen here.

Jodrell Bank. 'The observatory has taken a prominent part in the current exploration of the universe, including the approach to the discovery of quasars, and the exploitation of the pulsars' (p. 58).

'Thus Bell Labs launched a new kind of radio astronomy, concerned with cosmic background radiation' (p. 114).

*How cool is empty sky?*

If your television screen is specked with 'snow' – extraneous flashes in the picture – you are quite entitled to grumble about your fellow men who, with electrical apparatus of many kinds, contaminate your picture. But they are not exclusively to blame, because there is natural interference, too, coming partly from the Earth's atmosphere and partly from the Sun and the rest of the universe. Your television receiver is an inefficient kind of radio telescope. A certain amount of the 'snow' comes from radio galaxies and quasars. And some, though a very small part of it, comes from the Big Bang with which the universe began, when all the matter and energy was gathered in one place. At least, that is the present interpretation of a discovery surpassing even the quasars in its cosmological implications.

The Bell Telephone Laboratories in New Jersey boast a remarkable number of Nobel prizewinners among their former staff, showing that research in communications engineering touches on fundamental questions in physics. Bell Labs have also, quite accidentally, made two of the most important astronomical discoveries of this century. A telephone company is not, on the face of it, concerned about the stars. But it does worry about noisy lines – the telephonic equivalent of 'snow'.

In 1931-2, at Holmdel, New Jersey, Karl Jansky of Bell Labs was using a curious 'merry-go-round' aerial to investigate sources of

*First radio telescope.*
*A reconstruction (at Green Bank) of the aerial built by Karl Jansky nearly 40 years ago (at Holmdel), with which he unexpectedly detected radio noise coming from the universe. Note the wheels from a Model T car.*

noise in radio telephone circuits, when he discovered that some of the noise was due to radio waves coming from the Milky Way. This was the origin of radio astronomy. There was lack both of professional interest and of technology for pursuing Jansky's discovery, until after the second World War, but it led eventually to the current radio astronomy and a transformation of our view of the contents of the universe.

History repeated itself in a quite uncanny way, in 1965. By then, satellites were the big new thing in communications and Bell Labs were involved in the *Echo* and *Telstar* experiments. For this purpose, a steerable radio horn was built at Holmdel, about a mile from where Jansky's merry-go-round had stood. The horn and its associated equipment comprised a very sensitive receiver for microwaves – radio waves a few centimetres in wavelength. The detector itself, at the throat of the horn, was cooled with liquid helium to cut down noise originating in the receiver itself. While working with the horn, Arno Penzias and Robert Wilson found that weak but persistent microwave radiation came from the sky at all times of the day and night. Thus did Bell Labs launch a new kind of radio astronomy, concerned with cosmic background radiation not immediately attributable to particular objects in the universe.

Penzias and Wilson were very puzzled and mistrustful about their result, until they learned that at Princeton University, a couple of hours' drive away, the physicist Robert Dicke had predicted just such background radiation from the fireball of the Big Bang – supposing that our present universe came into existence in a great explosion, 10 billion years ago. A physicist at the University of Colorado, George Gamow, had predicted fireball radiation in 1948, but Dicke and his colleagues were preparing to look for the cosmic microwaves themselves, when the news of the Holmdel discovery reached them. Today, the Princeton group is still busy with special radio telescopes, studying the 'microwave background' – which they confidently label the fireball radiation. Other astronomers join in, with radio telescopes, optical telescopes and rockets, chiefly to confirm or disprove the true character of the background radiation.

Is it really the echo of the Big Bang, 10 billion years ago? The investigators use yet another, technically the most significant name for it, the '3K black-body radiation'. That rather curious term means the kind of radiation emitted by a perfectly colourless and matt material, at a temperature three degrees above absolute zero. It is important to the interpretation of the background radiation as due to the Big Bang, that it should be of this form.

Physicists can calculate exactly what the relative strengths of radio waves, heat waves, light and so on, should be at different wavelengths, for sources at different temperatures. If the fireball, with

which the universe supposedly began, radiated when it was at a temperature of a billion degrees, its strongest radiation would have been gamma-rays and X-rays, with a long tail of decreasing intensity, through the visible, infra-red and radio wavelengths. By now, according to the Princeton explanation, the fireball radiation has cooled, because of the expansion of the universe, to a very low temperature – '3K'. That means the greatest strength of this background radiation should be at a wavelength of one millimetre, with practically nothing at shorter wavelengths, but with a radio 'tail' detectable out to some tens of centimetres. There is no flexibility at all: if the 'background radiation' is really '3K black-body radiation' derived from the 'fireball radiation', precise numbers can be put on its present relative strength at each wavelength.

At the Bell Labs, at Princeton, and at Cambridge, England, radio measurements at different wavelengths have quickly established that the background does indeed conform extremely well to the predicted 3K pattern. The crucial test, though, comes at the peak of one-millimetre radiation, and beyond. Does the intensity rise and fall appropriately, at these wavelengths? Here life becomes difficult for the astronomer. The millimetre and sub-millimetre wavelengths fall in the awkward gap between radio and infra-red, which is the least accessible part of the whole electromagnetic spectrum. Not only are measuring techniques difficult, but the air strongly absorbs and emits radiation at less than three millimetres.

*Holmdel. This curious horn first picked up the microwave radiation from empty sky, which many astronomers interpret as the echo of the Big Bang with which the universe began.*

*Kandiah Shivanandan and colleagues. At the US Naval Research Laboratory an Indian-born rocket astronomer checks the telescope with which he has measured millimetre radiation from the sky.*

Astronomers have found two ways around this difficulty. The first is to use optical telescopes to read 'molecular thermometers' far away in space. Everything in the universe is presumed to be bathed in the background radiation, including the gas in the Milky Way which absorbs light from the stars. The effect of the background radiation is mildly to excite some of the gas molecules, so that they absorb energy of slightly longer wavelengths than they otherwise would. This effect is detectable in starlight. Indeed, a Canadian astronomer, Andrew McKellar, had spotted it 14 years before the background radiation was discovered at Holmdel; the explanation, though, was lacking. The trick is now revived, most notably by Patrick Thaddeus and John Clauser of the Institute of Space Studies, New York. Their findings indicate that the background radiation is at least roughly right at the shorter wavelengths, to fit the 3K pattern.

Otherwise, astronomers can measure the background at around one millimetre wavelength using telescopes despatched by rocket above the Earth's atmosphere. Martin Harwit and James Houck at Cornell University, and Kandiah Shivanandan of the Naval Research Laboratory, prepared 6½-inch infra-red telescopes for two short flights, carried out early in 1968. After prolonged tests on their equipment they announced that they have found radiation about thirty times as strong as would be expected from the 3K predictions. Quite apart from this discrepancy the result, if correct, makes the universe much warmer than anyone had supposed. As the total recording time from both flights is only five minutes, they cannot transform our view of the cosmos, or dispose of the fireball theory.

Further measurements are planned by these experimenters, and by others.

There is thus a clear contradiction between observations, at this turning point of the background radiation. The starlight studies rely on better-known techniques but are more indirect; the first results from rockets are suspect because the difficulties of direct observation at these wavelengths are notorious. In any case, confirmation of the strong radiation at one millimetre wavelengths will not, by itself, disprove the 3K pattern; there could be another source of radiation superimposed on the 3K background radiation. In the results from their first rocket flight, Houck and Harwit also find strong radiation in the infra-red, at wavelengths of about a tenth of a millimetre and below, and the possible explanations they put forward include radiation from dust lying between the planets or between the stars. Thaddeus, whose studies of starlight have tended to confirm the 3K pattern, believes that the rocket results must indicate some local source of radiation.

Nor, conversely, will confirmation of the 3K pattern necessarily prove that it really is the fireball radiation from the start of the universe. Opponents of the Big Bang theories hope, probably in vain, that the effects of very distant radio galaxies or infra-red galaxies may add up to provide the background radiation. Or there may be other mechanisms at work that keep the universe at an average temperature of 3K.

Nevertheless, the fireball explanation is the simplest and most convincing of any advanced so far, so that the background radiation constitutes the strongest evidence in favour of the Big Bang. It is not essential for the Big Bang story that such a background should exist: although any real Bang would have produced a great deal of radiation, the universe might have expanded enough, since the Bang, to render it quite undetectable. But the fact that the background does exist is doubly significant, not only as support for the Big Bang, but also as evidence that the origin of the universe may be still visible to us, in space and time. In other words, if this interpretation is correct, we have the whole history of the universe laid out for our investigation.

## Deductions in the dark

Darkness has connotations of evil, yet we should greatly miss the cooling and restful night if the Earth did not, once a day, spin our heads away from the Sun and towards the night sky. Investigation of the overall construction of the heavens (cosmology, for short) begins with the child's observation that the night sky is dark. It gives us surprisingly important information about the universe.

The simplest imaginable form of the universe consists of stars, or galaxies, dotted fairly evenly in an infinity of transparent space. Suppose we live in such a universe. Wherever we look in the night sky, we shall see a star. No matter how discriminating our telescopes, we cannot see any gaps between the stars because, along every line of sight, we eventually reach a star. The sky is certainly not dark at night; the whole of it, by night and day, is actually as bright and hot as the Sun and we are roasted out of existence.

We plainly do not live in a universe of that kind. Vast numbers of stars and galaxies are spread in a huge volume of space, yet we see so much blank space between them that the overall effect is dark. One possible reason is that the universe is not infinite in extent. Another is that the light from the most distant stars and galaxies fails to reach us. Either way, the result is similar: there is a horizon where the universe cuts off, or fades out.

This remarkably powerful argument about the darkness of the sky is normally called Olbers' Paradox, after an astronomer of Bremen who wrote about it in 1826, although Edmund Halley had propounded the problem a century before Heinrich Olbers did. Olbers offered a

*Olbers' Paradox. The universe cannot be infinite and entirely visible.*

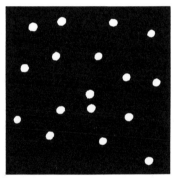

The sky is dark.

If there were a visible star at every point in the sky, it would all be as bright (and hot) as the Sun.

solution to the paradox of the dark sky. He suggested that the light from distant stars was cut off by intervening dust. Olbers was not stupid because obstruction does occur, especially in the disc of the Milky Way. He could not know that there is remarkably little obscuration of distant galaxies.

What other possibilities are there? The universe could be young so that, if we look too far, we see no stars because they did not exist in time to start sending light for us to see. That hypothesis does not seem too frivolous, either, in view of some current results. In fact, Olbers' Paradox was eventually resolved by another means: the universe is expanding, and that prevents the sky being light at night. Because distant objects are moving away from us at high speed, their light is greatly weakened. Dennis Sciama, the Cambridge cosmologist, calls Olbers' failure to think of this explanation, 100 years before the expansion was discovered, one of the greatest missed opportunities in the history of cosmology.

## Inside-out universe

Robert Dicke's group at Princeton are trying to measure our speed and course in relation to the universe as a whole. They are looking for slight variations in the strength of the microwave background radiation from different directions in the sky. The direction where it is strongest will, if they can find it, show which way we are heading. That the background is at least roughly the same everywhere is one of the reasons for ascribing it to a universal explosion. But the fact that they have so far failed to find any clear variation at all is becoming a little uncomfortable. For a start, it means that the fireball was remarkably round and uniform, with no lop-side burning. And if more precise measurements fail to reveal at least a slight difference in microwave intensity in the various directions, the implication will be that the Earth and our Galaxy are stationary in the centre of the universe! That idea, so obvious to ancient astronomers, would be quite incredible to their modern successors. They would sooner look for some quite different explanation of the microwave background. than presume theirs to be so peculiar a vantage point.

By now something may have struck the reader as odd. The microwave background is held to mark events at the original centre of the universe, and yet it comes from all directions, and from distances greater than any other we know. The universe seems to be turned inside out, with the fireball all around us. A simple-minded picture would have the fireball, if it were detectable at all, very small in one direction in space, at whatever distance we have travelled since we were ejected by the explosion. (By 'we' I mean the stuff of our

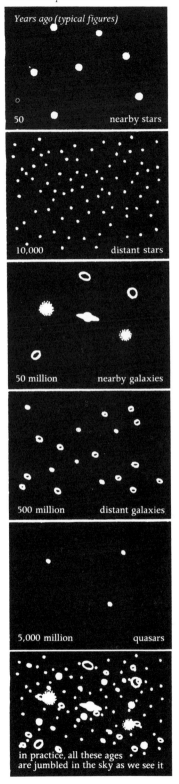

*The telescope as a time machine.*

Years ago (typical figures)

50      nearby stars

10,000      distant stars

50 million      nearby galaxies

500 million      distant galaxies

5,000 million      quasars

in practice, all these ages are jumbled in the sky as we see it

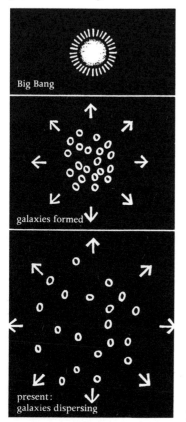

*Big Bang theory. The matter of the universe expands and disperses from a creative origin.*

Big Bang

galaxies formed

present:
galaxies dispersing

Galaxy.) A brief answer is that, according to the Big Bang theory, we were *inside* the fireball, and the whole fireball has expanded around us: the universe *is* the fireball, cooled down.

We cannot always believe our eyes, or trust to common sense. Just as an underwater swimmer has a very peculiar view of the world above the surface, so for men contemplating the universe the image is greatly distorted – despite the transparency of space. The effects can be weird and astronomers have to try to allow for them, in interpreting what they record in their telescopes.

*Distortions because information is obsolescent.* Although it is very interesting that our telescopes show us distant galaxies, not as they are now, but as they were billions of years ago, the picture is not the same as what we should see if light travelled instantaneously. The galaxies certainly would be much farther away now; their speeds probably would have changed, too; and the most distant galaxies themselves would be seen to have evolved and changed. To add to the confusion, nearby, distant and very distant events are superimposed, in our picture of the sky, as if it had been exposed over and over again by an incompetent cameramen.

*Distortions because the universe is expanding.* Measurements of the recession of the distant galaxies give the same illusion as the uniformity of the microwave background – that we are at a still, central point of the universe and everything else is rushing away from us. It is another kind of optical illusion; astronomers in any other galaxy would have just the same impression. The reason is not hard to grasp: all the distances between all widely-spaced galaxies are increasing. Perhaps the rate of the expansion of the universe is slowing down, in which case another odd effect may occur: galaxies previously invisible can come into view, because their redshift diminishes even though they are still moving rapidly away from us.

*Distortions because light paths are bent.* Light from one galaxy on its way to another (ours, for example) is subject to the gravitational pull of the universe as a whole. If the universe originated from some centre, the distribution of matter in space and time, seen from a point away from the centre, will be lopsided, and a ray of light will bend towards the centre of gravity, as if space were a lens. If the universe is sufficiently massive and compact, light cannot escape from it into any remote regions beyond the galaxies.

Reverting to the Princeton group's attempt to fix our trajectory in relation to the universe as a whole, we find David Wilkinson and Bruce Partridge, physicists in Dicke's department, continuing several years' work. They are seeking some slight variation of the microwave background. 'We should at least be feeling a breeze due to the Earth's motion,' Wilkinson points out. Currently they are using a pair of horn detectors on Princeton's Forrestal Campus, to look for any systematic change in the strength of the microwave background in directions and different times. During an earlier experiment on White Mountain, California, the Princeton physicists found no variation greater than one part in a thousand. They are trying again, at a longer wavelength and with greater precision. One difficulty Wilkinson explains sadly: 'It takes us a year to get a number.'

## A narrow choice

Two mistakes of cosmic dimensions helped to stimulate 20th-century cosmology. The first was by Albert Einstein. Soon after he developed his new theory of gravity, during the first World War, he tried to deduce from it the general nature of the universe. He came to the conclusion that the universe must be rigid and unchanging. In fact he had made an error of algebra and a philosophical oversight. Alexander Friedman in Russia saw that the theory also admitted the possibility of an expanding or a contracting universe. Subsequently the discovery by Edwin Hubble and his American colleagues that the universe did indeed seem to be expanding gave a great fillip to the theory, and the Belgian priest Georges Lemaître capped it with the idea of a primaeval Atom and the Big Bang. Subsequently, George Gamow and Robert Dicke were to produce refined versions of the Big Bang theory, the latter's invoking a theory of gravity different from Einstein's.

A second fruitful error was observational rather than theoretical and was made by no less an observer than Hubble himself. He underestimated the distances of the galaxies by so large a factor that, when their movements were traced back to a common centre, it looked as if the universe was much younger than the Earth! This absurdity provoked Hermann Bondi, Thomas Gold and Fred Hoyle to consider how a universe could be both expanding and eternal. Their answer was the Steady State theories of 1948, developed and published independently by Bondi and Gold on the one hand and Hoyle on the other. They allowed that the galaxies would move apart, but held that the appearance of the universe could remain more or less unchanged, if new matter were created continuously to fill the gaps left by the older galaxies moving apart.

By the time Walter Baade at Palomar in 1952 had discovered that Hubble's result should be corrected, making the universe older than the Earth again, the Steady State theories had shown enough youthful vigour of their own to remain an alternative view of the universe even when the Big Bang was no longer ridiculous.

When astronomers express dissatisfaction with both the Big Bang and the Steady State concepts of the universe, they are in trouble,

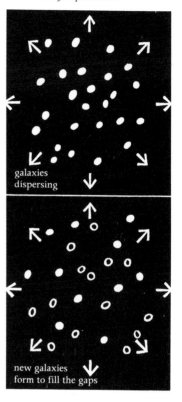

*Steady State theory. The matter of the universe, though expanding, is continuously replenished.*

galaxies
dispersing

new galaxies
form to fill the gaps

because it is hard to imagine radical alternatives. If you accept the strong evidence of the telescopes, that the contents of the universe are flying apart, and have been doing so for billions of years, the obvious deduction is that those contents were formerly much more closely crowded together than they are now. In other words you have the universe of the Big Bang. Various accounts are possible of the nature and history of the explosion but, whatever the details, the galaxies are flying apart like shrapnel from a bomb. It implies drastic change in the appearance of the universe and also a singular moment of creation, before which our universe did not exist as we know it.

The only way anyone has thought of, to avoid this conclusion about the former condition of the expanding universe, is to suppose that there was no such crowding, because less matter existed in the universe than does now. That matter is being created continuously is the essence of the Steady State theories. The proposal of continuous creation, which contradicts long-held scientific and theological beliefs, is not made any cheekier by requiring the rate of creation of new galaxies to be just sufficient to compensate for the expansion, so that the universe looks much the same at all times. In the Steady State theories the universe is infinite in extent and infinitely old. The whole thing is expanding. The galaxies we see have all been created, one by one, in the vacuum left by older galaxies moving apart. We see only a small sample of the universe, according to the Steady State theories, while, if a Big Bang theory is right, we can see a substantial part of the whole universe.

The early versions of the Steady State theories supposed that matter was created uniformly throughout space – it was the easiest idea to

*Cambridge, England. The one-mile radio telescope consists of three dishes, one of which is movable. When observations are married by computer, the picture of radio sources so obtained is equivalent in detail to what would be provided by one huge dish.*

deal with mathematically. But now Hoyle thinks it takes place in localised sources. Moreover, in their revised Steady State theory, Hoyle and J. V. Narlikar allow for variations in the rate of input of matter. If the rate increases, the expansion of the universe speeds up; if the rate diminishes, so does the expansion rate. In other words, you can have an unsteady Steady State, but one which will always settle down to a steadier condition unless the rate of input of matter changes rapidly. By such variants Hoyle and Narlikar allow themselves bubbles of different densities within an infinite universe.

## Census of explosions

To visit the one-mile telescope at Cambridge on an ordinary day can be an uncanny experience. Here is one of the most important scientific instruments in the world, which continuously hunts for new radio sources and maps known ones in detail. And yet there is no one around. Even when you enter the control room, it is deserted, with only a logbook lying on a table to show that humans were here. A computer, not at all unlike the bizarre confections of the science-fiction moviemakers, runs the show, with flashing lights, chattering tape readers, and intermittent hoots of greeting.

With that infinite patience better demanded of electronics than of scientists, the computer keeps three 60-foot dishes gazing fixedly at one spot in the sky. The dishes swing as slowly as the hour hand of a clock, to compensate for the Earth's rotation. Two of the dishes are on fixed mountings half a mile apart; the third is on a railway half a mile long, running farther on from one of the fixed dishes so that, at the limit, the outside dishes are a mile apart. When the radio astronomers want to survey an area of the sky, they set the telescope to watch the same area day after day. But each day they shift the movable dish one step along its track. As the Earth rotates, the line of dishes swings round.

By this procedure, the telescope amasses vast records, but these make no sense until they are combined in Cambridge University's Titan computer. Then – and this is the sleight of hand perfected by Sir Martin Ryle and his colleagues – the results printed out correspond to pictures of the sky such as would be obtained with a radio dish a mile in diameter! And this combination of telescopes and computers is the instrument with which Ryle and his colleagues detect radio galaxies and quasars 100 times fainter than any they have picked up before.

Since the early years of their science, when the powerful radio source Cygnus A turned out to be a distant galaxy, many radio astronomers have thought they hold the aces in the game of discover-

*Exploding Universe*

*Sir Martin Ryle. For two decades he has led the distinguished group of radio astronomers at Cambridge.*

ing the overall nature of the universe. A good deal of disagreement has raged for years among radio astronomers, and between radio astronomers and theorists, on what the correct observations of distant radio sources amount to, and what they say about cosmology. But there is little doubt remaining, in 1969, about the general sense of the message from the faintest objects the radio telescopes have picked up. The universe has changed.

Some radio astronomers, particularly Ryle, go much further than that bald statement. The one-mile telescope at Cambridge already seems to give a rough picture of the complete universe because it accounts for most of the detectable radio noise from the sky, leaving no room for any substantial addition to the furniture of the universe.

The discovery of the microwave background, suggestive of the Big Bang, was quick and accidental. The premeditated and much more time-consuming way in which radio astronomers have sought to discover the overall nature of the universe is to survey the sky as a whole and count the number of sources they discover. Sharp controversy has added spice to the history of these efforts. Between Ryle's group in England and Bernard Mills's in Australia there was competition to get results and disagreement about the actual measurements and counts. The Cambridge radio astronomers made, at an early stage, embarrassing errors. They published a list of 2000 radio sources, of which only a quarter were correct; many of the rest were inventions of the particular type of radio telescope employed! They quickly redeemed their reputations with a series of catalogues (3C, 4C and 5C) which are part of the bedrock of current astronomy – witness the frequent reference in this book to 3C48, 3C120, 3C273 and so on, numbers in the third Cambridge catalogue.

There was also public dispute, most memorably between Ryle and Fred Hoyle, about the interpretation of the results of the surveys. Hoyle being a man who does not give up the Steady State easily, we need not be surprised to find that dispute still simmering in 1969, even though the British and Australian results now largely concur about the facts of the radio universe.

The radio astronomers set out to see what the universe was like at different ages in the past. If the universe is everywhere much the same, in all regions and at all times, as in the basic Steady State theory, then the radio sources will be scattered in approximately equal density at all distances. But if the universe is less uniform, or if it evolves (changes with time), then the density of radio sources will vary at different distances. In other words, there will be a discrepancy between the numbers of strong, nearby sources, and of faint, distant ones. In particular, according to the Big Bang theory of the exploding universe, the galaxies were much more crowded together at their formation than they are now. If typical faint radio sources lie far away, and therefore far back in time, they will be progressively more numerous at each increment in distance than the Steady State theory predicts.

When it comes actually to detecting and counting the faint radio sources, the picture is complicated by the expectation in both kinds of theory, that the number of *observable* objects in a certain volume of space will actually decrease at great distances. The reason is the redshift which, when it is large, represents so great a loss of energy that many sources must become undetectable. Nevertheless, the Big Bang theory still predicts *relatively* more faint sources than does the Steady State theory.

The latest radio counts show that large-scale variations exist in the universe, which are devastating to the simple Steady State theory. But the details are even more interesting. So far from the density of faint sources decreasing, it actually increases out to great distances. The pattern does not correspond with the simple picture of an exploding universe, in which galaxies like those we know nearby sprang ready-made from a Zeus-like centre. There seems to be a period of infancy in the universe, eight or nine billion years ago, when exploding galaxies and quasars were much more numerous, or more powerful, than they are today.

What does Hoyle say to all this? He concedes that the relatively nearby region of the universe is not uniform, and he and his colleagues modify the theory to allow for local variations from place to place in the infinite universe of the Steady State. But he disputes the conclusions from the faint and supposedly disturbed sources, on the grounds that the radio astronomers do not know that they are counting. Few of the sources are identified with visible objects and the quasars; if many of them are in fact quasars, the questions are still open, about what quasars are, and how far away they are.

And how does Ryle rebut Hoyle? The 'local' variation Hoyle invokes would have to extend over the whole of the observable universe, and even then we should have to be in a very special position in the universe – a situation, Ryle remarks, 'distasteful to astronomers since the time of Copernicus'. And as for doubts about the quasars, if they are strongly represented among the unidentified radio sources then they must be powerful objects at great distances.

Hoyle has an ally on the other side of the world. John Bolton was a radar officer in the Royal Navy who found himself ashore in Australia at the end of the war and became one of the founding fathers of radio astronomy, *primus inter pares* in a brilliantly successful group at Sydney. He played an outstanding part in the identification of the main classes of radio sources – supernova remnants, radio galaxies and quasars. He has also made it his business to become a skilled optical astronomer, so that he can use the big optical telescopes himself for the investigation of quasars. He is in charge of the Parkes 210-foot radio telescope – the Australian equivalent of Jodrell Bank – and, among other things, Bolton's radio astronomers use this big dish to hunt for quasars.

'We are finding quasars almost at will,' Bolton says. He disagrees with Ryle in the interpretation of the counts of faint radio sources. He finds more difference between radio galaxies and quasars than the Cambridge group does. Bolton also thinks that what faint sources are picked up towards the limits of a radio telescope depends so much upon what radio wavelength is being used, that counts of faint sources have to be treated with great scepticism.

To the long-standing observer of the little world of astronomers, it will seem that a lot of the sport has gone when the two antagonists, Ryle and Hoyle, stop disagreeing. They must do so some day, because from just such controversy, so conducive to bright ideas and yet so intolerant of error, does science arrive at durable conclusions.

### Stellar archaeology

The Royal Greenwich Observatory is no longer by the Thames at Greenwich. Successors of those Astronomers Royal who made Greenwich Mean Time and the Greenwich longitude the reference standards for the world found the street-lights and smoke of London increasingly frustrating for their work. The move was made in 1957 to the 15th-century moated castle at Herstmonceux in Sussex. In 1968 a new giant telescope came into operation.

Bernard Pagel, of the observatory's staff, uses the 98-inch Isaac

Newton telescope to examine the chemical elements present in very old stars. The elements reveal themselves by their characteristic wavelengths in the light of the stars. Pagel is anxious, as are other astronomers, to deduce what raw materials were available when these ancient stars were formed. They were plainly different in composition from those out of which the Sun and the planets were made, much later. Detailed chemical analyses of the old stars may therefore give important information about the early history of our own Galaxy and of the universe as a whole.

If the original matter of the universe was the commonest element, hydrogen, theorists can see three ways in which the heavier elements can have been formed. The first, chronologically, is in a Big Bang at the start of the universe. Calculations say clearly that a good deal of helium, but very little of anything else, could be produced in that tremendous explosion – assuming it really occurred. The second possibility is in the explosion of a huge mass in the centre of the Galaxy, very soon after its nativity. Depending on circumstances before and during the explosion, it could have manufactured a variety of elements, though not all of them, and distributed them for incorporation in the stars then forming. Finally there are the explosions in ordinary big stars, which supply every known element for later stars. Supernovae make the elements in the right abundances to explain the composition of the Sun and the planets.

The very oldest stars travel in disorderly orbits outside the disc of

*Herstmonceux Castle, Sussex, built of brick in the 15th Century, became the headquarters of the Royal Greenwich Observatory in 1957.*

*The chief instrument at Herstmonceux is the 98-inch Isaac Newton telescope, the largest optical telescope in Western Europe. Its dome stands 90 feet high.*

the Milky Way. They contain far less in the way of heavy metal atoms than do the oldest stars in the disc of the Galaxy. In fact, in order to account for the apparently sudden manufacture of heavy metals, in a generation of exploding stars, the Galaxy should have been blazing 100 times more brightly than it does now. If the oldest stars contained absolutely no heavy metals there would be no problem about the earliest stages of the Galaxy. But their sparse inventory requires explanation. Did it come from a central explosion in the Galaxy or a quick succession of early supernovae? The relative abundances of various elements depend on which mechanism was responsible for their formation.

Pagel's present view is that there is no strong evidence on which to invoke a special event, such as a galactic explosion, to explain the heavy elements detected in old stars. But he continues his patient inquiry – a kind of stellar archaeology – to reduce the uncertainties and settle arguments. Even with a 98-inch telescope, several hours' exposure of a particular star may be necessary to obtain the spectrum which reveals the details of its composition.

But what about the helium, supposedly made in the Big Bang? In the Steady State theory the raw material of the universe, produced by continuous creation, is thought of as being purely hydrogen, and all the other elements are made in subsequent nuclear processes in stars and galaxies. But the Big Bang, as a gigantic H-bomb, would have enriched the whole universe with helium. Jesse Greenstein of Palomar has found old stars remarkably deficient in helium. That looks bad for the Big Bang theories, but Greenstein says there is no way of being sure that the helium is not buried deep inside the star, where it cannot be seen. He sees no prospect for deciding cosmological issues by this test, in the near future.

## Birth of the galaxies

Both Maarten Schmidt, optical astronomer, and Sir Martin Ryle, radio astronomer, believe they are seeing to the limits of the universe of galaxies. Schmidt thinks so because, despite a lot of looking by himself and others, they cannot find quasars with a reddening of their light corresponding to a distance greater than about 8·5 billion light-years. He wishes he could do so, because that would help to deflate the arguments of those who still contend that the distances ascribed to quasars are mistaken and they are really nearby. The reason Schmidt gives for this failure to find more distant quasars is that there are none to be seen. He is looking so early in the history of the universe that it is too young to have started making quasars.

As for Ryle, his source counts show, as mentioned, a great number of powerful radio sources at a distance of 7 or 8 billion light-years. But, further out, the number of sources begins to fall, and then so rapidly that Ryle deduces that his one-mile radio telescope has reached back to a time when there are no sources to be seen because they do not exist. He doubts whether much new information can come from more sensitive radio telescopes, because these will pick up a large number of intrinsically weak radio sources that are relatively nearby, and little from farther away.

*Birth of the galaxies. According to Cambridge radio astronomers, there is a horizon of time before which radio galaxies and quasars did not exist.*

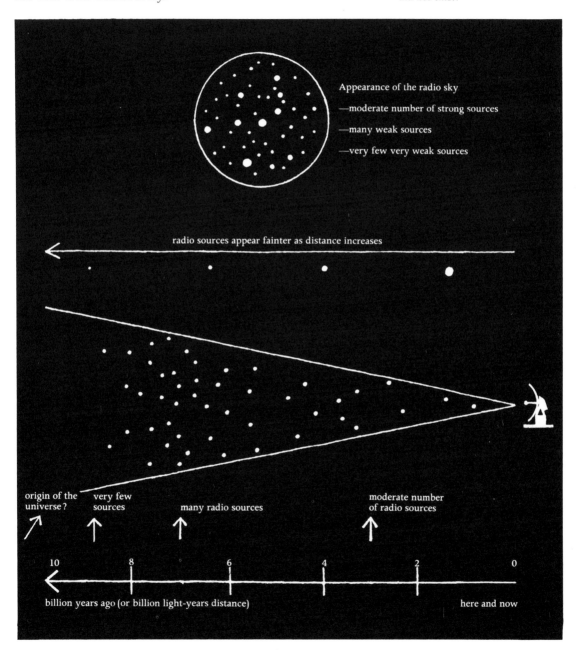

Appearance of the radio sky

—moderate number of strong sources

—many weak sources

—very few very weak sources

radio sources appear fainter as distance increases

origin of the universe?    very few sources    many radio sources    moderate number of radio sources

10    8    6    4    2    0

billion years ago (or billion light-years distance)    here and now

Attempts to understand the history and general nature of the universe become ensnarled, at this stage, with questions about the origin and early history of galaxies, and the nature of the exploding galaxies and quasars. The only help in disentangling the simultaneous effects of the evolution of the universe as a whole and the evolution of the radio sources comes from measurements of the total radio energy from the sky – which sets a limit to what else might be there. M. S. Longair, in Ryle's group, roughs out a general story of radio sources in the early stages of the universe. It respects the limit set by the total radio energy and also fits the Cambridge source counts.

Longair's story can be put into simple terms as follows. The maternity ward of the universe, where the galaxies are forming, is about 9 billion light-years away, or 9 billion years *ago*. Beyond and before that natal region there is nothing to record except the microwave radiation from the supposed fireball that marked the origin of the universe itself. But when the galaxies do come into existence, during a short period of time, there is a great commotion; among them are many quasars and radio galaxies – many more very powerful ones than we see among nearer and more recent objects. The population of inherently weak objects is not, however, unusual at that early period. Thereafter, and coming to closer objects, the number of very powerful sources drop off very rapidly, as if the disorders of quasars and radio galaxies become less likely as the universe matures.

The most redshifted quasar known, at the time of writing, is Margaret Burbidge's. The leading woman astronomer of our time is the daughter of a chemist; she learned her craft in the little observatory of London, at Mill Hill, before going to Chicago and California to work with the big American telescopes. Nowadays she is professor of astronomy on the San Diego campus of the University of California, a leading authority on the quasars and one of the few lady Fellows of the Royal Society. The quasar 5C2.56, which Margaret Burbidge finds to have a redshift of 238 per cent, would normally be called the remotest visible object known to man (or woman). But as her husband, Geoffrey Burbidge, is the outspoken champion of the view that quasars are closer than most people think, the appellation might be tactless.

### Not with a whimper

Allan Sandage, astronomer at Palomar, says that the heavens will fall within 70 billion years. He makes careful measurements of the speeds with which the galaxies are flying apart, and he concludes that there is a slowing down, as if the far-flung galaxies, like stones thrown in the air, will eventually fall back on their tracks. The present expansion

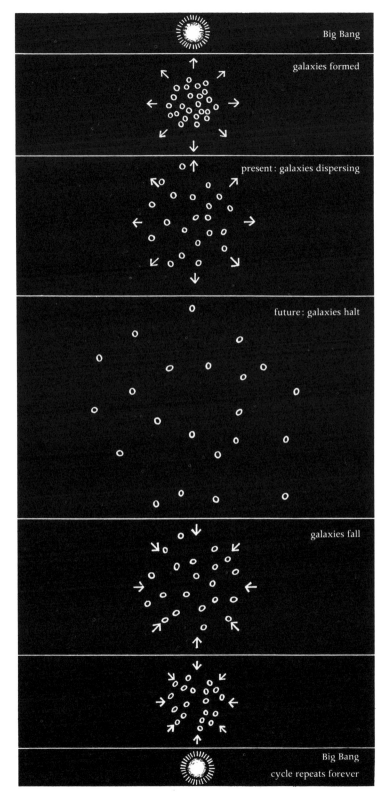

Big Bang

galaxies formed

present : galaxies dispersing

future : galaxies halt

galaxies fall

Big Bang
cycle repeats forever

*Oscillating Universe theory. The stuff of the universe, scattered by one Big Bang, eventually stops travelling outwards and falls back together, causing another Big Bang from which a new universe can be made.*

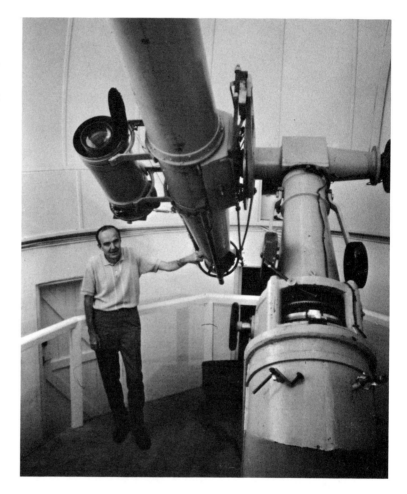

*Allan Sandage. The distinguished American astronomer from Palomar visits Australia in 1968–9, and uses one of the small instruments on Mt Stromlo to photograph the nearby Magellanic Clouds. He thereby obtains pictures equivalent in detail to those of the more distant galaxy M33 seen in the 200-inch telescope.*

of the universe will then give way to a contraction. Eventually all the matter in the universe will crash together and, in Sandage's view, the great mass so created will explode in its turn, bringing a fresh universe into being, phoenix-like out of the ashes of the old. Similarly, the Big Bang that marked the start of our universe also proclaimed the demolition of a preceding universe.

Such a story of an explosive cycle of death and re-birth, for which Sandage now believes he has evidence, is known as the Oscillating Universe. It envisages changes throughout the life of a universe, as with the one-shot Big Bang, and yet it provides a kind of eternity. Sandage estimates that the life-span of our universe is 80 billion years, and that we are one-eighth of the way through it. In other words, from an explosion 10 billion years ago the universe will go on expanding for another 30 billion years and then spend 40 billion years collapsing.

Sandage's determination of the slowing down of the universal expansion is a factor that has to be taken into account in any attempt to date the origin of our universe from the motions of galaxies. If the aim is to work backwards from the distances and speeds of the galaxies, to find how long ago they may all have been closely packed together, it is important to know whether their speeds have changed since the beginning of the universe. Some slowing down is to be expected, unless one imagines the existence of some special force that keeps driving the galaxies apart against their mutual gravitational forces. From his measurements of the speeds of galaxies, Sandage finds evidence that these were indeed faster in the past. It is from the rate of slowing down that Sandage calculates a standstill in 30 billion years, with collapse to follow.

Robert Dicke, who predicted the microwave background as an echo of the Big Bang, is another adherent of the theory of an Oscillating Universe. Indeed, his interest in the microwave background radiation, before its discovery, stemmed from the cosmological problem of how to make hydrogen at the start of our universe. While physical processes are known on Earth, which create matter out of energy, these invariably create 'anti-matter' as well, the stuff that reacts with ordinary matter in mutual annihilation. The apparent absence of this explosive partner from the substance of our universe creates a fundamental difficulty for a one-shot theory of the Big Bang. But if the matter of our universe is scrap from an older universe, that pushes the problem of the existence of matter into a distant past. For Dicke, the fireball of the Big Bang is not just an act of creation: it is an incinerator in which the material of the last universe was broken down into hydrogen, before being scattered again to form the galaxies of our universe.

Dicke's description of the course of events in the fireball makes use of the modified theory of gravity which he had earlier developed with Carl Brans. As he describes it, the fireball had a temperature of more than 10 billion degrees and a diameter of only a few light-years, at the instant marking the extinction of the previous universe and the start of our own. Within four minutes the temperature fell sharply to one billion degrees and the beginning of the expansion enlarged the fireball to about 80 light years. Six minutes later the figures were one hundred million degrees and 800 light-years. But the radiation observed as the microwave background today was emitted from the fireball at the earlier stage, when the temperature was one billion degrees. Since that time the universe has grown three hundred million times bigger and the radiation three hundred million times cooler – to three degrees (3K).

There are snags in the idea of the Oscillating Universe. For a hundred years physicists have believed, on very strong evidence, that the universe must always tend to become increasingly disordered. Whatever its overall architecture may be the principle in question is the second law of thermodynamics which says the universe must grow tepid. Yet an Oscillating Universe is reheated and reorganised; it seems to have a fresh start every 80 billion years or so, when the accumulated debt of disorder is cancelled.

An observational snag for the Oscillating Universe is that there is not, at first sight, enough matter in the universe to bring about the collapse that Sandage and Dicke envisage. The gravitational force exerted on a galaxy by the rest of the universe seems to be too low to arrest its headlong flight. Each galaxy will lose speed, certainly, but the dispersal of the galaxies reduces the gravitational force still more, and the galaxies will go on flying apart for ever. Such is the conclusion to be drawn if most of the matter of the universe is visible to us in the form of galaxies of stars. The density of visible matter in the universe is too low, by a factor of about 100, to bring about its eventual collapse under gravity. The hunt is on, for dark or transparent matter in the universe that could contribute sufficient gravity to lasso the galaxies and corral them for another Big Bang.

*Invisible masses*

Jan Oort of Leiden is an astronomer whose contributions and influence are exceptional among his generation. He and his group are best known for their work on the structure of our Galaxy, including the fulfilment of H. C. van de Hulst's prediction that the characteristic radio waves emitted by hydrogen gas would allow the mapping of the Galaxy (see page 48). Currently, the radio astronomers at Leiden, under Oort's leadership, are using their radio telescope at Dwingeloo to study clouds of gas travelling at speeds of up to about 100 miles a second, towards our Galaxy from outside. These clouds were first detected in 1964 and, on the basis of careful investigation since then, Oort concludes that these clouds have slowed down since encountering the outlying gas of the Galaxy. They were originally coming even faster. The clouds vary in mass, but are typically around 10,000 times the mass of the Sun. Oort calculates backwards from the present observations to the time when the Galaxy and the gas clouds were formed, and thereby computes that the mass density of the universe is not quite sufficient to stop the expansion of the universe from continuing indefinitely. The additional matter is ten times the mass of the conspicuous galaxies.

The detection at Dwingeloo of infalling gas is one of the few signs of matter in the universe besides what is immediately conspicuous in the galaxies. The vast spaces between the galaxies are extremely transparent to light and radio waves – otherwise we should not be able to see as far as we do in space and time. Nevertheless, it is possible that a vast amount of hot, electrified hydrogen gas fills intergalactic space.

Richard Henry of the Naval Research Laboratory in Washington believes he has detected this invisible plasma and that it is sufficiently dense to close the universe and bring about its eventual collapse. Lest 'dense' gives the wrong impression, I should add that this hydrogen comprises no more than ten atoms per cubic yard of space. Nevertheless, it amounts to 100 times the mass of all the galaxies.

The first indication of the existence of this vast addition to the substance of the universe came in a rocket flight of an X-ray detector in September 1967. A background of low-energy X-rays coming from the universe was detected and interpreted as emission from hydrogen gas at a temperature of 800,000 degrees. Henry and his colleagues, and Minoru Oda of Tokyo University, have been repeating the rocket experiment early in 1969.

Other invisible forms of matter in the universe include the scarcely detectable neutrinos. We must certainly live in an ocean of neutrinos but estimates of its possible density make it an unimportant contributor to the collapse of the universe. There are also, within galaxies, planets and burnt-out stars that we cannot see because they are cool – but these are fully accounted for in overall estimates of the mass of a galaxy. Yet another possible contribution to the gravity of the universe is particularly bizarre: very massive entities that have disappeared out of our universe, as it were through holes in the sky, leaving behind only their gravitational fields, like the grins of so many Cheshire Cats.

*Richard Henry, rocket astronomer at the Naval Research Laboratory, Washington DC. Henry was responsible for the flight that detected diffuse hydrogen between the galaxies.*

## Bully for the Buddha

'Sire, I had no need for *that* hypothesis!' Pierre-Simon Laplace retorted, when Napoleon asked why his great treatise on the mechanics of the universe contained no reference to the Almighty. Yet only by stopping short can one discuss the nature of the universe without meeting the age-old question of whether or not the hand of God is perceptible in the cosmos. Even if the answer is an irritated 'No!' it ought to be sought. Not four centuries has elapsed since professional astronomers believed that heaven, complete with angels, existed up in the sky, and that only 'our muddy vesture of decay' prevented us hearing the music of the seraphs.

In the Far East, as an historian of astronomy tells me in Tokyo, there is really little interest in the origin of the universe, because of the pervasive Buddhist belief in cyclical patterns of events. In Israel, I meet a leading physicist who wonders why there are so few Jewish astronomers, even after Albert Einstein's contributions to universal issues (in fact, several distinguished theorists are at work, though few in Israel). And, as Sir Bernard Lovell reports with dismay, the official Marxist-Leninist philosophers, who proscribed the translation of Western books dealing with the evolution and expansion of the universe, have also restrained Russian astronomers from working on the ultimate cosmological problem. It is the more ironical, seeing how Alexander Friedman's name figures in all those foreign textbooks, as a pioneer theorist of the expanding universe. For some reason the cosmological debate is most compelling to men brought up on both the Old and New Testaments, whether they believe them or not.

The questions scientists ask and hypotheses they entertain are influenced by philosophical attitudes of their society, but their discoveries can be iconoclastic. Especially in research into the origins of things – of the universe, of the Earth, of life and of man – the scientist intrudes into the domain traditionally proper to the Divine Creator. There is a well-known formula for a stand-off: science deals with facts, religion with faith and never the twain shall meet. It is a simplification that saves a good deal of argument in the still hours of the night. Scientists rarely claim expertise in theology, and churchmen have learned by hard experience not to do so in science. When they picked fights with Galileo and Darwin, they were bound to come off worse. The historical trend is nevertheless unmistakable. Science has progressivly eroded the area in which divine intervention is necessary or even admissible. Religions that once offered to explain everything, and claimed commonplace natural phenomena as deliberate acts of God, are now confined, in their mundane scope, to those areas where the scientists are still only groping for understanding – particularly the workings of the human mind. Even there, the prospect is plain enough: before the end of the century we should know in detail how the brain works, how we think and why we feel.

The evidence grows that everything after the creation of our Galaxy – including the origin of the Earth and of life – are explicable as a chancy but not mysterious series of physical and chemical processes. Any opportunity for supernatural explanations of the material world is therefore driven right back to the creation of the matter of the universe – to be literal, out to at least 8,000 million light-years from here, where we lose sight of events. Is there a constructional job for God, so far away, so long ago?

Monseigneur Lemaître certainly thought so up to his death in 1966. In propounding his theory the primaeval Atom, the Belgian astropriest explicitly sought to modernise the opening verses of the Bible without contradicting them. Yet, if theology is to play the cosmic game at all, the contrary view, that of continuous creation, would seem better matched to the vision of an immanent Deity. If the third view is correct, and the universe goes through an endless cycle of

*Arecibo. The aerial platform of the 1000-foot radio telescope in Puerto Rico hangs 500 feet above the reflecting bowl in the valley. On the catwalk leading to the platform, shown above, the author is accompanied by Frank Drake.*

*Parkes. Australia's equivalent of Jodrell Bank, in the outback of New South Wales, has a fine 210-foot radio telescope as its main instrument.*

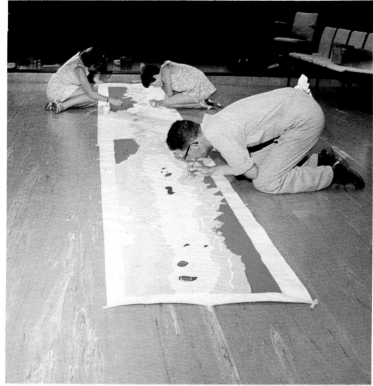

*Painting a map, already outlined by computer, which shows radio emissions from the spiral arms of our Galaxy, the Milky Way, as recorded with the big dish at Parkes.*

*Kitt Peak. Filming the BBC-PBL programme on astronomy on the tower of the solar telescope. In the mirror can be seen a reflection of Baboquivari, the nearby mountain that Papago Indian legends call the centre of the universe.*

*Frank Low and Lear jet. Low flies an infra-red telescope at 50,000 feet above the water vapour of the atmosphere, to observe 'warm' galaxies.*

explosion, collapse and explosion, creation as such can be pushed infinitely far back in time – bad for Moses but bully for the Buddha.

Although I have not evaded the big question I would not presume to give a definite answer; I would only suggest that, so long as religious folk presume a workaday operational connection between God and the physical world, their beliefs will continue to be vulnerable to the discoveries of science.

### The first know-alls

This very time of ours, seven-tenths of the way through the 20th Century, is unique in the long history of the endeavours of men to understand the universe they inhabit. There never was a period like it, and there is never likely to be such a climax again. One reason is the opening, in the past twenty years, of the wide range of new windows on the universe:

electromagnetic waves: radio waves
infra-red rays
ultra-violet rays
X-rays
gamma-rays
matter: primary cosmic rays
neutrinos

To say we can see 'everything' can no longer be much of an exaggeration. Barring the discovery of completely new forms of energy or matter, our children and even our great-great-grandchildren will have few new windows to open. The chief exceptions are hypothetical: low-energy neutrinos and gravity radiation may pervade space, but we have no good means of detecting them yet. Otherwise, the nets are out, to catch anything that can bring us information about the universe. In some cases, as for sub-millimetre waves, gamma-rays and high-energy neutrinos, the nets are so far rudimentary. But they exist in our era, and we shall probably witness whatever big discoveries they or improved versions of them can make.

Men of our time can see to the bounds of our universal parish. At long last we descry that edge of the world, off which old-time seamen feared to fall. No ship could get so far. Never mind, in this connection, whether the precise interpretations of what we see and cannot see are correct. The awesome realisation is that we have reached a point, with radio observations of faint sources, beyond which there is nothing to see, and never can be. Perhaps we see no farther because, as Ryle says, there is nothing there – because nothing but radiation exists when our time-machine telescopes take us back to the newborn

*Opposite:*
*Arno Penzias (in front) and Robert Wilson of Bell Telephone Laboratories, who discovered the mysterious 'microwave background' from the universe during experiments connected with early communications satellites.*

universe. Or the universe really does extend much farther – perhaps it is infinite – but the relentless redshift erases all information at a horizon where objects seem to travel away from us at the speed of light. Whatever the reason, we have encompassed all the universe we can know. We can expect the view to change appreciably only if we watch for many millions of years.

That is not to say big discoveries like the quasars and pulsars are no longer possible: indeed, they become more, not less likely as techniques of observation improve. There will be no end to the filling in of detail in the map of the observable universe. Neither will there be lack of substance for controversy about the essential character of the universe. But the boundaries of the argument, like the boundaries of the universe itself, are fixed by features already apparent to us, at least in outline. Can our generation, the first know-alls, make approximate sense of it all? Or will our descendants smirk about our ideas as we do about those of our ancestors?

### Cosmic stocktaking

George Gamow of the University of Colorado, one of the chief architects of the Big Bang theory, was a good-humoured fellow and until his death in 1968, he liked to cite a poll of 33 leading astronomers, taken by Science Service in Washington, ten years ago.

|  | *Yes* | *No* | *No comment* |
|---|---|---|---|
| Did the universe start with a Big Bang? | 11 | 12 | 10 |
| Is a poll of this kind helpful to science? | 0 | 33 | 0 |

All the hard work astronomers are doing just now, to discover what they can about the nature of the universe, would be pointless if it could be settled by a vote, or if the outcome of present observations could be predicted. Yet, before long, there will surely be a clear majority of experts in favour of a particular cosmology. As doubts abate, and critics grow weary, then cosmology will be incorporated into the body of accepted scientific knowledge. It will then be taught to students and infants, and will become part of the general knowledge of the human species. The 'official' cosmology is likely to harden, for a while, into dogma. Although, like all scientific knowledge, it will still be vulnerable to amendment in the light of new observations or ideas, whatever cosmological conclusions astronomers come to, within the next two or three years, are likely to persist for a generation or more.

Nor is it hard to guess what the cosmology is most likely to be. A judicious view is that of Philip Morrison, of MIT. He personally

detests the Big Bang theories, and would much prefer the universe to be in a Steady State, but he is forced to admit, as a physicist, that the evidence is piling up in favour of the Big Bang.

Unless reworked Steady State theories make a strong comeback in the early 1970s, or someone quickly thinks of a quite original cosmology, the Big Bang will come to have almost universal acceptance (no pun intended) as a simple, all-embracing description of the cosmos we inhabit. Who knows, perhaps it is correct?

The only foreseeable development that can halt the present inexorable swing in favour of the Big Bang is strong evidence that the theory of gravity is wrong. And there, in the view of some defenders of the Steady State theories, lies the real significance of the quasars.

*Philip Morrison of MIT. Visiting Kitt Peak, and standing by a mural illustrating ancient Mayan astronomy, he arbitrates between the Big Bang and Steady State theories for the BBC-PBL astronomy programme.*

*Haystack. 'A group at the Massachusetts Institute of Technology repeats, with the most modern aids, one of the historic tests that encouraged physicists to admit that Einstein could be right' (p. 146).*

# Chapter IV: New Laws for Old

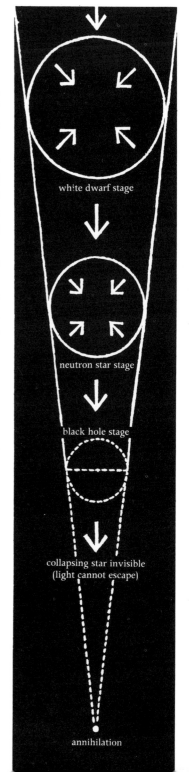

*Gravitational collapse of the central remains of a star, as predicted by current gravitational theory. In this diagram, the objects are not to scale.*

white dwarf stage

neutron star stage

black hole stage

collapsing star invisible
(light cannot escape)

annihilation

## Black hole of gravity

In Chapter I, astronomers were watching the force of gravity at work, as it conjured stars and planets out of clouds of gas, ignited the stars' nuclear fuel and, when that fuel ran out, crushed the stars into white dwarfs or pulsars. The same force was somehow, in Chapter II, aggregating vast amounts of matter at the centres of galaxies, to bring about the cataclysms we observe as quasars or exploding galaxies. In Chapter III, gravity was threatening to make the heavens fall, in an Oscillating Universe doomed to consume itself, so that a new universe can spring from the fireball. Even though that last hypothesis is still in dispute, gravity plainly belies the sober connotations of its name. There is nothing dull about gravity: rather it is like gravidity, pregnant with creation.

Gravity is capable of very strange tricks indeed if the predictions are to be believed, which Roger Penrose of London University and Kip Thorne of Caltech make. Hints of these appeared earlier (page 92) when astronomers were fretting about the durability of quasars, or the possibility of disappearing stars. The point is that any massive object should eventually vanish through a hole in the sky. This bizarre idea is a direct consequence of the law of gravity, as at present understood.

When matter reaches a sufficient density, so the story goes, the force of gravity becomes so strong that it crushes the matter out of existence, or at least out of our universe. Gravitational collapse of a cloud of gas is necessary for the birth of a star, but the collapse is brought to a halt by the heat and radiation generated at the centre. When, much later, the star runs out of nuclear fuel, it collapses further, to become a white dwarf or a neutron star. But, if the residue of the star exceeds a certain mass, gravity can overwhelm even the nuclear forces that hold up a neutron star, and make the collapse catastrophic and total. The more closely gravity confines the material the stronger the gravity grows and no other known force of nature can resist it. The star is snuffed out and in its place there is a black hole of gravity.

Astronomers would have to be very lucky, and very sharp-eyed, to see a star doing its disappearing trick. As Thorne describes it, if the collapsed core of a big star is only twice the mass of the Sun, it can survive neither as a white dwarf nor even as a neutron star. It just goes on collapsing. To an outside observer, the star will be seen to shrink, in a matter of seconds, until it is about four miles in radius. Although the star itself goes on collapsing, faster and faster, the astronomer outside would see an eight-mile disc simply go red and then fade after a few more seconds. The resulting black hole persists indefinitely, and any unfortunate space travellers who happened to stray across the four-mile limit would be swallowed up and disappear as if into a bottomless pit. A bigger mass would collapse a little more slowly, in days rather than seconds, and it would leave a wider black hole.

What happens to the matter that has vanished? In some cases it may be crushed to zero size. In other cases it can survive after a fashion, even though it has certainly departed from our universe. It may constitute a mini-universe of its own. Or it may reappear in another, quite different universe. Or it may pop up, at some completely different place and time, in our own universe. None of these bizarre possibilities, reminiscent of the time-travellers of science-fiction, is ruled out by Einstein's theory: indeed, Yuval Ne'eman of Tel Aviv University, and other theorists, have suggested that quasars are matter from massive objects, exploding back into the universe after a catastrophic collapse.

If you find this idea of total collapse mystifying and incredible you

*Wallace Sargent, a young British*
*astronomer now working at*
*Palomar, contributes to the*
*BBC-PBL programme.*

have allies in Fred Hoyle and other theorists who say that a law of gravity predicting the existence of black holes in the sky must be wrong. Such theorists repudiate both the black hole and, with them, the current theory of gravity. If the black holes, or gravitational 'singularities' as the astronomers call them, are to be forbidden, there is no easy way out of the difficulty. Einstein's law of gravity cannot be redeemed by adding a clause. You cannot, for example, say that the gravitational force becomes a repulsion at very short ranges, or that some other activity of matter produces a repulsion that stops the collapse. In Einstein's theory, any forces of such kinds in a highly collapsed star will only make the situation worse.

The laws of nature proclaimed by scientists must be at least approximations to reality or they would not survive scrutiny for long. Eventually, however, observations of greater precision, or else newly discovered tricks of nature, may cast doubt on the accepted ideas. Scientific laws are not immutable but they are rather precise and specific. Even small and local discrepancies can force a complete overhaul, and an accompanying revolution in ideas.

### Another source of energy?

Eta Carinae is a charming name for one of the most beautiful features of the southern sky, a strangely coloured collection of objects visible only to telescopes in the Southern Hemisphere. 'Long-playing supernova' may be the aptest way to describe it; though whether this is an accurate description remains to be seen. Astronomers have watched it, fascinated, for more than a hundred years. In 1843, Eta Carinae, previously an unimpressive star, burned as bright as Sirius. It blazed for fifteen years and then grew much dimmer again, becoming invisible to the naked eye by 1867.

Recently it has begun to grow bright again. But interest in 1969 stems chiefly from Gerry Neugebauer's discovery that it is an extraordinarily powerful radiator of infra-red rays – the brightest object in the whole sky at a wavelength of a fiftieth of a millimetre. Like the pulsars, Eta Carinae is a small-scale mystery, to set alongside the much more massive quasars as a phenomenon not easily explicable by current ideas.

'Clearly there's something very odd going on here. Either we need a new source of energy or the familiar sources are working in ways that we haven't yet discovered. The Sun generates energy by thermonuclear reaction, but this source isn't powerful enough for the quasars. We have to think again.'

The speaker is Wallace Sargent, commenting on the theoretical problems thrown up by the discovery of the extraordinarily compact, extraordinarily bright objects far away in the universe. As Sargent says, nuclear forces seem inadequate to sustain the outpouring of energy from the quasars. The most obvious alternative source of energy is the gravitational force. In a massive object or a dense crowd of smaller objects, collapse is plainly a means of supplying a great deal of energy in a small volume. Whether or not they are the source of energy, the gravitational fields are immense in a quasar. But then another question crops up at once: why does the quasar not continue to collapse, catastrophically, and snuff itself out?

Similar problems flow from the pulsars. Although these are much weaker objects than the quasars, they too represent condensed matter, of dubious viability according to present theories. There are also, in the pulsars, concentrations of broadcast energy not easy to explain.

As gravity is the best hope for explaining the energy of the quasars and pulsars, a revised gravitational law may show how previously unsuspected transformations of energy can occur in extreme conditions. That is, on the whole, a more probable outcome than the discovery of some completely new force in the universe, ranking with gravity, electromagnetism and nuclear forces. Such a mechanism may also help to explain other events, less striking only by degree, such as the fierce light and heat at the centre of the long-playing quasars of the seyfert galaxy, or the long-playing supernova of Eta Carinae. Indeed, few of the dramatic occurrences in the universe, whether the birth of stars, their extinction as supernovae, or the formation and transformation of galaxies, would escape the influence of modified gravitational law.

There is no need to exaggerate the extent of the uncertainty. A great deal of what we know about stars and galaxies is not in question. Most of the furniture of the universe is still the familiar, enduring stuff; the objects that command attention are abnormal, but there is a great deal of normality around. In the early days of astronomy, interest focused on a handful of planets, because they moved among apparently fixed stars. Today the strong and peculiar sources of radio noise clamour for attention. Like doctors concerned with the sick rather than the healthy, the astronomers cannot help but be preoccupied with the pathology of the universe.

## Checking up on Einstein

Astronomers so anxiously mind their p's and q's, their pulsars and quasars, chiefly because these objects seem to show gravity with intensity never observed before. Under extreme conditions, slight flaws in known scientific laws can be magnified and therefore become discoverable.

*Sir Isaac Newton and Albert Einstein. They laid down the theory of gravitation which is now questioned by some astronomers.*

Our current theory of gravity we owe to Albert Einstein, who drafted it half a century ago. His General Relativity tells much the same story as Isaac Newton's theory, but it predicts different effects in special situations. 'Such a beautiful theory!' the late Robert Oppenheimer once remarked to Robert Dicke, speaking of General Relativity. 'It's a shame there are no experiments.' Oppenheimer exaggerated only mildly. Aesthetically, the theory is unequalled in the history of science, but the tests made in the past 50 years, to check the theory, are scrappy and unimpressive, when you think of its fundamental importance. In effects that physicists and astronomers can observe at reasonably close quarters, the difference between the predictions of Newton's, Einstein's, and other theories such as Robert Dicke's of Princeton, are very slight, yet translated to universe-wide operation these subtleties become all-important. Fortunately, several new ways of checking Einstein's theory have become available, just when they are most needed at this critical stage of astronomy – though whether experiments performed in the gravitationally calm environment of the Earth and the solar system can teach us anything about the laws prevailing inside a quasar remains an open question.

A group at the Massachusetts Institute of Technology repeats, with the most modern aids, one of the historic tests that encouraged physicists to admit that Einstein could be right. Photographs taken during an eclipse of the Sun in 1919 revealed that the apparent positions of the stars beyond the Sun were slightly altered, by the bending of their rays as they grazed through the gravitational field of the Sun. The question was by how much? The answer was that the stars were shifted in apparent position to just the extent predicted by Einstein. But then, and since, the measurements have not been precise enough to be wholly convincing. The MIT group is repeating the experiment, without needing to wait for an eclipse. They use radio telescopes far apart to measure how the apparent separation of two convenient quasars (3C273 and 3C279) changes as the Earth's motion brings the Sun nearly into line with them. They employ the technique described earlier (see pp. 102ff) for jointly operating two radio telescopes – in this case at Green Bank and Haystack – separated on a long baseline. It relies on simultaneous recording of radio signals at the two places, using atomic hydrogen clocks to synchronise the tapes, which are then compared by computer. It is one of the most precise astronomical techniques available in 1969. At the time of writing, the experiment has been done, but the results have not yet been fully extracted from the 700 tapes collected in 15 days of observation. A parallel experiment is measuring the separation of the two telescopes to within a few inches in 500 miles.

Tracking the planets by radar is potentially one of the most powerful new ways of checking on Einstein. An earlier success of General

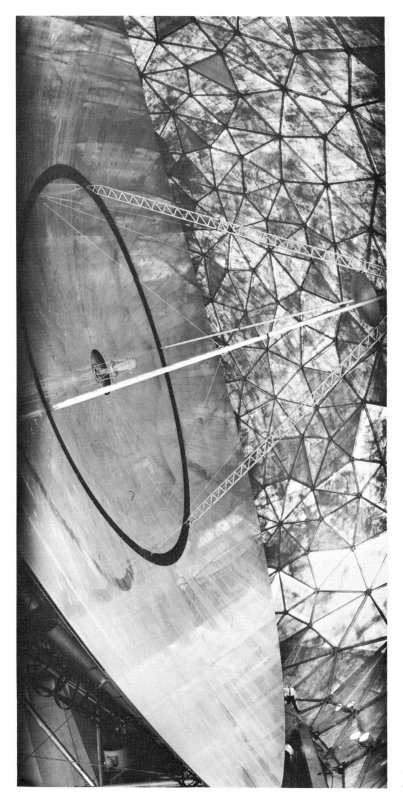

*The Haystack radio telescope of the
MIT Lincoln Laboratory has a
120-foot dish with a very
accurately figured surface.*

*The Haystack telescope was built in 1964, inside a big weatherproof 'radome', which is transparent to radio waves.*

Relativity was to explain why Mercury's orbit gradually drifts round the Sun; now American radio astronomers at Haystack and Arecibo expect to be able to measure the rate of drift very accurately by radar, and also to see whether the orbits of other planets, Earth and Mars, show the same effect. But that is inherently a very slow test of the theory of gravitation, because the change in Mercury's orbit has to be watched for ten or twenty years.

Quicker and more direct is a check on another prediction, from Einstein's theory, that radio waves will slow down as they pass close to the Sun. If a planet, being tracked by radar, moves on to the far side of the Sun, the radar echoes from the planet should be slightly delayed. Irwin Shapiro, Gordon Pettengill and their colleagues at MIT have carried out such experiments, and their first results showed that the predictions of General Relativity were at least roughly borne out. Pulsars also offer a way of checking the same effect, because the time of arrival of the pulses, which can be predicted very accurately, should be slightly delayed when a pulsar and the Sun are nearly in line, as seen from the Earth.

Satellites provide yet more possibilities for gravity experiments. For example, one of the experiments being prepared in the USA envisages the launching of a very accurate atomic clock into orbit round the Earth. If Einstein is right, the clock should run slightly fast, when it is in orbit, to the extent of gaining a thousandth of a second every three weeks, compared with an identical clock on the ground.

## 'Prove Fred wrong'

Fred Hoyle is the most celebrated theoretical astronomer of our time. He holds the same professorship at Cambridge as did the late Arthur Eddington, the man who showed how mathematics could penetrate into the very hearts of the Sun and stars. At his Institute of Theoretical Astronomy, Hoyle seizes on all the new information coming from the observatories and spins hypotheses around it. He infuriates his fellow astronomers by frequently changing his mind, and by saying things that turn out to be invalid. Yet, more than any other theorist, he has stirred the pot of astronomical ideas and a great deal of productive observational work has been inspired by the wish to 'prove Fred wrong'. Sometimes he is disconcertingly right, as when, visiting the California Institute of Technology, he predicted the properties of a particular atomic nucleus from general considerations about processes in stars; American physicists thereupon made the necessary measurement and found it to be exactly what Hoyle had said. But Hoyle's work is not just a series of squibs. For twenty years he has questioned the most fundamental notions of current science and now he senses the approach of a great denouement, in which new physics will come out of astronomy.

*Fred Hoyle. Leading protagonist of the Steady State theory, he now directs the Institute of Theoretical Astronomy, Cambridge.*

Hoyle and his young Indian colleague, Jayant Vishnu Narlikar, seek effects of distant matter on local events, and connections between the grandest manifestations of matter – the universe at large – and the properties of the smallest sub-atomic particles. They venture to relate gravity and the expansion of the universe to the creation of new matter, within the updated Steady State theory. They develop a theory of electromagnetism in which the nature of the universe (in Steady State, of course!) enters in a fundamental way in the interactions of electric particles. But Hoyle is after bigger game still.

Perhaps there is no need to wait for refined experiments before passing judgement on Einstein's theory of gravity. For Hoyle the very existence and persistence of the quasars is proof enough that ideas about gravity are wrong. Otherwise these objects should simply vanish into black holes – singularities. Talking to his colleagues at the Royal Astronomical Society in London, about the difficulties of explaining sources of energy like quasars, radio galaxies and seyfert galaxies, Hoyle declares: 'The behaviour of the sources cannot in any case be explained in terms of conventional physical theory. According to conventional theory, large masses in small volumes plunge into singularities . . . I would expect the sources to fade away . . . in a time scale of the order of a year. Instead we observe violent outbursts in a time scale of the order of a year. The observed properties are exactly opposite to what we would expect according to conventional physical theory.'

Hoyle reasons that, if the theory of gravitational collapse is wrong, so is the theory of the emergence of the universe from the greatest singularity of all – the Big Bang. But for him that is only a debating point; the important task is to find the new physics which explains what we observe in the sky. And, as for the direction in which the evidence is pointing, Hoyle thinks matter behaves differently in highly condensed objects from the way it does in more ordinary regions of the universe, like our own, where there is no specially strong force competing with the influence of the cosmos as a whole.

### The new astrology

Have we mere by-laws in our part of the universe – local ordinances of nature valid on Earth and perhaps throughout our own supercluster of galaxies, yet with no universal rigour? The scientists who read the fine print of nature here on Earth, believing they are on the track of universal facts, may be deceiving themselves. There may be a sense in which their knowledge is temporary and parochial. Even with the recent great enlargement of our view of the universe, our sentiments should still be those of Newton, who saw himself as only a boy playing on the seashore, 'whilst the great ocean of truth lay all undiscovered before me'.

Astrologers nowadays use computers to cast horoscopes. Their continuing claims to foretell the fate of human individuals, from the influence and relative motions of the Sun and planets, are at best idiotic, at worst fraudulent. It is just as absurd to suppose that the location of objects millions of miles away, at the time of your birth, fixed your temperament and fortune for life, as it was to regard the comets, as the ancients did, as harbingers of disaster. As there may be other, much subtler and more credible ways, in which distant objects in the sky can influence our circumstances, it is as well to anticipate 'I told you so's' from the astrologers, and make it plain that we are talking about something quite different from the personalised nonsense in which they trade.

For those who see no point in spending taxpayers' money on telescopes, because what goes on in distant galaxies is of negligible significance to us on Earth, a variety of answers can be offered. Among them are discoveries of great relevance to earthbound science and engineering, like Newton's laws of motion and gravity, the element helium first detected in the Sun, and the thermonuclear process of the stars. Nobody has yet confirmed or refuted, except in debating terms, the more philosophical proposition of Ernst Mach a century ago that the distant matter of the universe has a profound effect on conditions on Earth. Some distinguished commentators have found the idea ridiculous – a modified hocus pocus like that of the astrologers. Among those who have taken it very seriously indeed was Albert Einstein, who was deeply influenced by Mach's notion that the mass of a body was generated by the effect of the other matter in the universe. The efforts of Hoyle and others, to relate the behaviour of electric particles to the nature of the universe as a whole, were mentioned in the previous section.

If there is really some sense in which the strength of natural forces depends on what is going on elsewhere, those forces may vary from time to time or from place to place. The obvious working assumption is that the same natural laws applied everywhere and at all times. The fact that the same characteristic wavelengths can be seen in the light both of laboratory lamps and of distant galaxies suggests regularity in the universe. But other things may 'give'.

Some scientists have speculated that the force of gravity may have been stronger in the past than it is today. Yet the evidence of long, calm evolution of life on this planet argues strongly against any such change, at least locally. If gravity were stronger, the Sun would have burned much more brightly than it does today because the tighter packing would have raised its temperature. The Earth would have been scorched, the oceans would have boiled, and life of the terrestrial kind would have been impossible.

Another guess was made by George Gamow that the electric charges of atomic particles may have been less in the past than they are today. Such a change would mean, among other things, that the light emitted by atoms would be weaker – redshifted, in fact. But again, scientists can find local arguments against it. One effect would have been to speed up radioactive decay in some atomic nuclei in the past; the fact that the nuclei in question can still be found in the rocks of the Earth makes Gamow's idea unlikely.

With the quasars, another possibility arises: variation in laws from place to place. A strictly local violation of well-known physical laws may occur where matter and energy are packed very tightly. One suggestion has been that electric charge, which in our world always comes in multiples of a fixed quantity (the charge of an electron), may come in fractions in the quasars, as a result of the formation of new types of fundamental particles in the extremely violent conditions in those objects.

It might be comforting if familiar ideas could be saved in this fashion, by shuffling off the violations to untypical places and treating them as additions, rather than alterations, to the natural laws. Unfortunately any such intention conflicts with the broader ambition of science – that of finding general theories covering all circumstances. More probably, the natural design of the universe is so interconnected in all its parts that even the distant quasars reflect upon our own circumstances and natural laws.

Variations from place to place may nevertheless occur in the universe, and may have important consequences. The successful origin of life on this planet depended on a scarcity of oxygen. If there had been too much oxygen around, the complex chemicals of incipient life would have 'burned up'. The abundance of oxygen made, like other elements, inside stars, is limited by the fact that another nucleus, beryllium-8, is not stable. If this nucleus, which goes towards making oxygen, were only slightly more stable, the resulting excess of oxygen would have been fatal on the early Earth. If regional variations in the laws of physics in the universe should cause beryllium-8 to be more stable in some places, there may be huge populations of apparently normal galaxies in which life as we know it is simply impossible.

### Watch this space

Neither the big questions nor the big discoveries of the astronomers are very difficult to grasp, but my account has been misleading to the extent that it has skirted technical considerations in the interests of simplicity – making everything seem easier than it really is. Also I plead guilty to emphasising dramatic discoveries of astronomers, without dwelling on their longer-term investigations which consolidate their knowledge and which check the elaborate calculations of the theorists against the facts of the universe. The expertise of the astronomers is as formidable as their jargon. Unless individuals were prepared by years of study and experience, against a background of centuries of star-gazing and the perfection of technique, the big discoveries would not have been made. If getting important results is a matter of luck, it is strange that some astronomers are consistently 'luckier' than others.

The discovery of quasars flowed from a century's investigation of the light of stars and galaxies, on the one hand, and from fifteen years' careful study of radio sources in the sky, on the other. Though the discovery was unexpected, the techniques that made it possible were not acquired by chance. Maarten Schmidt's flash of inspiration, when he recognised the large redshift of 3C273 for what it was, could only have come to someone who understood the technique and theory thoroughly, who knew all the reasons why the redshift should not

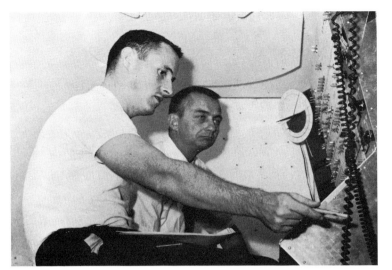

*Towards space observatories. One of the uses to be made of experience and space vehicles acquired for manned flight to the Moon will be the construction of manned observatories in orbit, where astronomers will be able to study the universe by rays that do not penetrate to the ground. Left, American scientist-astronauts practise with a mock-up of the proposed Apollo Telescope Mounting (ATM). Below is a telescope prepared by American Science and Engineering for the ATM. Below left, more ambitious ideas for the future include this larger observatory in space.*

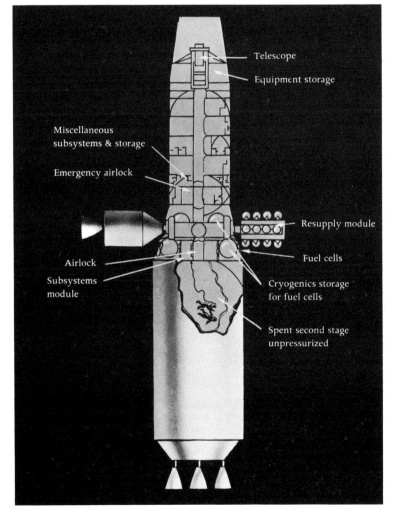

Telescope

Equipment storage

Miscellaneous
subsystems & storage

Emergency airlock

Resupply module

Airlock

Fuel cells

Subsystems
module

Cryogenics storage
for fuel cells

Spent second stage
unpressurized

be true, and yet could be convinced that it was true.

Much the same is the case for the discoveries of the microwave background radiation and of pulsars. Men had been working with microwaves for more than 20 years, unaware of the background. Unless Penzias and Wilson had perfected equipment of rare sensitivity they would not have spotted it. Similarly, the equipment with which the pulsars were discovered was unusual and ingenious – actually designed for hunting quasars! – but even then the deep knowledge and quick wits of Anthony Hewish and his colleagues were necessary to recognise the first pulsar and to show that the obvious explanation – man-made interference – was wrong.

The story of astronomy will never be finished, the files closed, or the observatories scrapped. New techniques of observation and theory will clarify our view of the contents of space. Astronomers will go into orbit, or to the Moon, for a better look at the universe. Ground-based instruments will be made more sensitive and precise. Bright youngsters learning the astronomer's craft will bring fresh ideas to the subject. But, when tales of the cosmos are told, this period of ours may always be recalled as that in which men first came to realise what a violent universe they inhabit.

*After quasars and pulsars, what?
The exploration continues with
new radio telescopes, like this
French one at Nançay, south of
Paris. Its focussing reflector (left)
is a vertical cliff of lattice-work
1000 feet long. A tilting reflector
(right) directs radio waves from
the sky on to the vertical reflector.*

# Index

M